SEA BATTLES IN MINIATURE

A Guide to Naval Wargaming

SEA BATTLES IN MINIATURE
A Guide to Naval Wargaming

Paul Hague

*This book is published and distributed
in the United States by the*
NAVAL INSTITUTE PRESS
Annapolis, Maryland 21402

Patrick Stephens, Cambridge

First published—1980

Acknowledgements

The author would like to thank Brian Monaghan for taking all the photographs used in this book. He would also like to thank Martin Priddy, Bob Swan and Vince Driver for redrawing the diagrams and tables.

British Library Cataloguing in Publication Data

Hague, Paul
   Sea battles in miniature.
   1. War games, Naval
   I. Title
   793'.9   V250

   ISBN 0 85059 414 6

Text photoset in 10 on 11 pt English Times by Manuset Limited, Baldock, Herts. Printed in Great Britain on 100 gsm Matt Coated Cartridge, and bound by, The Pitman Press, Bath, for the publishers, Patrick Stephens Limited, Bar Hill, Cambridge, CB3 8EL, England.

Contents

Introduction

In the early 1960s the hobby of wargaming really took off in a big way, with the result that what was regarded 20 years ago as a crackpot pastime for the eccentric few has become established as a respectable and accepted hobby, with a following of tens of thousands. All this time, though, the emphasis has been on land wargaming and to the great majority of people, cognoscenti and non-wargamers alike, the hobby conjures up mental pictures of crowds of model soldiers fighting the battles of Alexander or Napoleon across a table top. Very few give thought to the naval side of the hobby.

This discrimination against the minority area of the hobby is unfortunate and it is difficult to see why there are not at least as many sea battles as land games, particularly in this country with its long naval association. Battle games using model ships are in every way as enjoyable as those using model armies and, because the psychological elements of morale and reaction are of less significance with ships than with men, they are capable of attaining a much greater measure of realism and accuracy. Not least of the virtues of naval wargaming is cost; a wargames fleet costs only a fraction of the price of an army of metal figures and can be painted and sent steaming into battle in a matter of days rather than months.

My wargaming both by land and sea has about a 15-year history and it has always comprised weekly battles against one or two regular opponents, so that the rules given in this book have the characteristics of home rather than club wargaming. They have no pretensions to being standard or definitive, having been developed for the particular requirements of our own campaigns, wars and battles, but they have proved themselves over the years, giving wargames that are fast-moving, exciting and realistic. No rules, however, be they ever so long, will legislate fairly for every possible situation which may arise in a wargame and so it must be taken as implicit that the players show a spirit of friendly compromise and deal with freak occurrences in a fair manner to ensure a historically accurate result. Those players who employ gamesmanship rather than tactics and seek out loop-holes in the rules rather than military opportunities are not worth playing against.

Paul Hague
Sale, Cheshire, October 1979

Chapter 1

About naval wargaming

When H.G. Wells' *Little Wars* was published 65 years ago it not only sowed the seed that has grown into the modern hobby of wargaming, it also established the tradition that such books should commence with a potted history of the hobby. Starting with its speculative beginnings in the mouth of a Neanderthal cave, writers since that time have progressed warily through the legendary tales of Uncle Toby and Corporal Trim, galloped up to the firmer ground of the Kriegspiel, Stevenson and Wells, and then rested their pens awhile, a tricky job well done, on reaching the late 1950s, the dawn of the modern age. The truth of the matter is that the authenticated history of wargaming, and by that term I mean the simulating of battles and wars in a more or less realistic manner for *pleasure,* is really quite short: about 100 years or so.

Wargamers, as much as anyone, like to see their roots deeply embedded in the past and this yearning for 'establishment' often leads to excesses of enthusiasm when trying to identify the founding father of the hobby. One of my wargaming opponents has decided in his own mind that the military models in his tomb show Tutankhamun to have been the world's first wargamer whilst another friend, with a hagiological turn of mind, has investigated the lives and deaths of certain obscure early Christian martyrs in order to find a suitable patron saint. For my own part I incline to the wet blanket opinion that wargaming as a hobby that we would recognise does not go back beyond Robert Louis Stevenson's parlour floor wars of the 1880s, which coincided with the availability of suitable model soldiers at a low price.

Naval wargaming has always been something of a Cinderella branch of the hobby and this is reflected in the tales of its legendary origins; there are none. No rumours of cavemen, or Uncle Tobys and Leading Seamen Trim or of literary giants, exist to tantalise the historian of naval wargaming, it just seems to have sprung out of nothing at the turn of the century when Fred T. Jane, the editor of *Jane's Fighting Ships* invented his naval Kriegspiel.

Although the game was primarily intended for use by naval officers it could also be used by amateurish civilians seeking an enjoyable pastime rather than an instructive exercise because, unlike the Kriegspiel used by the army at this time, the rules were not over-complicated and they allowed the game to work itself rather than requiring umpires to decide the effect of firing. The game's suitability for use as a hobby was further enhanced by the fact that it could be purchased in boxed sets, each set comprising rules, a dozen or more model ships, damage record cards and strikers (the tools which simulated gunfire).

These sets came in various sizes but the cheapest was priced at three guineas. As that sum in 1900 would have represented about a fortnight's wage for a working man it is perhaps not surprising that the game's popularity amongst civilians did not spread like wildfire.

In the 1930s in America the hobby seems to have had a short-lived golden age. A military writer called Fletcher Pratt invented a naval wargame of great ingenuity and realism; the rules were so simple and easily understood that members of the public with no previous experience of the game were recruited to command battleships and cruisers and destroyer flotillas in battlegames of reputedly enormous size that took place in public ballrooms. The war seems to have killed off the public's enthusiasm for such belligerent games and it was not until the 1960s, when wargaming in general really took hold as a popular pastime, that the naval wargame finally came to stay.

Since the early 1960s the number of naval wargamers has increased umpteen-fold, but as we were in the minority to begin with we are in the minority still with, at a guess, perhaps one sea-dog for every 15 or so landlubbers. Why this should be so is something of a mystery to me as wargaming is tied up with an interest in military history and, in this country at least, our naval history is so very much more creditable than our martial performances on land that one would have expected that the principal wargaming interest and activity would have been in assembling scaled-down fleets of 'wooden walls' rather than armies of the Peninsula.

A possible reason for the lesser popularity of naval wargaming is that whereas a land battle can be set up on the table top, fought out by the players and then the armies packed away and the battle forgotten about, the naval game does not so easily lend itself to the one-off battle. When the General sits astride his horse and plans the battle his primary concern is with the terrain. Is that marsh sufficiently impassable to rest his flank upon? How would that stream affect an attack on the right wing? Can that wood be used to prevent the enemy striking at his communications? Thus the tactics which an army adopts are as much determined by where it is fighting as what it is fighting about. The Admiral, on the other hand, is faced by a flat expanse of water, he commands a force of ships which are far more difficult to replace than men and the very fact that he is at sea suggest that he has some objective. His tactics will be dictated to a far greater degree by strategic considerations; he is sure of destroying the enemy but can he risk damage to his battleships which may put them out of action for a long period? The enemy cruiser force is temporarily stronger than his but if he does not force an action they will lose themselves in the vast expanse of ocean and prey on friendly commerce; can the strong enemy battle squadron be kept occupied sufficiently long for his frigates to attack the transports which they are convoying? If our naval wargame is to have any meaning, therefore, it must have a before, to determine the tactics which the fleets will adopt, and an after so that the real victor can be distinguished from the player who simply sinks the most ships.

One easy way of putting a sea battle into the larger context of a war is to invent a situation and base the criteria of victory upon it: 'You are a dirty dago in command of these galleons; to the north, out of sight, are your transports carrying soldiers to Bergen-op-Zoom to oppress the valiant Dutch; I am a gallant English sea-dog in command of this squadron, and if I manage to beat you and sail off to the north by night-fall I am able to attack your convoy and you have lost, if you prevent me leaving the fighting zone before night-fall you have won.' Such a verbal scenario will often result in a good battle, with both sides acting according to the strategic requirements and with a basis for awarding the laurels in the case of an incomplete result. All too frequently, though, things will go awry; those two bravoes Sir Horatio Hague and Le Comte Hanley de Knutsford, his inveterate opponent, have been known on many occasions completely to forget the whys and wherefores within minutes of smelling powder and on such occasions the battle invariably ends with the two Admirals, their fleets reduced to splinters, arguing over who won by least ignoring his objectives.

By far the best way of obtaining realistic results from naval wargames is to fight them as part of a campaign. This involves making a map of an area of sea and devising a strategic situation which gives each side a fair chance of winning. Units of the fleet are moved about the map (a number of methods of moving secretly are possible) and only when they meet are the model ships actually laid out and a battle fought. In this way we can bring in all the important factors which are inevitably left out of one-off battles, fuel, ammunition supplies, repair time, forces over the horizon, etc; not only do we get battles with a real purpose but also a method by which tactical defeats can be offset against strategic victories. The methodical, skilful player is vindicated and the *beau sabreur* put in his place. In addition to the much greater realism of the actual battles we also get the opportunity of practising grand strategy—we are elevated

As the Channel Fleet rides at anchor von Hanley's torpedo boat flotilla steams out of the dawn in a surprise attack. The Great Naval War of 1898 resulted in the surrender of all Britain's overseas possessions to France and Germany, yet it is little known to historians!

The Battle of Blenheim set up for a wargame. Unlike refights of historic naval actions, the terrain features instantly identify a land wargame.

in status from mere Fleet Commander to First Sea Lord or even Prime Minister.

By using campaign maps we also overcome the other main drawback of naval wargames—the inability satisfactorily to re-create a historic battle. If the land gamer wishes to refight, say, Waterloo he will set up a table-top battlefield featuring all the terrain characteristics of the original and he will deploy upon it model armies scaled down from those which actually took part. When he comes to fight, the terrain features will influence the thinking of the two players in the same way as they influenced Napoleon and Wellington so that the likelihood is that similar tactics will be employed and even if the wrong side wins the presence of the Château of Hougoment and the farm house of La Haie Sainte, etc, will distinguish the battle as Waterloo. If we are refighting Trafalgar, however, it is quite a different matter. Water looks much the same the world over and if the two commanders decide upon different tactics from Nelson and Villeneuve the battle may as well be a re-creation of a fight between Russians and Swedes in the Baltic for all the resemblance it will bear to Trafalgar. To re-create a historic sea battle, then, we must change our ideas about the meaning of the word 're-create'—we are not simply going to refight the battle, we are going to re-create *the circumstances that led to its being fought*. If we are going to fight Trafalgar by this system we will need to provide ourselves with not only two scaled-down fleets but also a map of the campaign area showing the West Indies in the west and the Levant in the east, Africa in the south and the South of England in the north. All the relevant fleets are in position in their cruising grounds or in port and Villeneuve may move his fleet where so ever he wants providing that he achieves the ultimate objective of covering the movement of the *Grand Armée* across the Channel, or perhaps, so as to bring in a proper degree of the 'fog of war', of invading Ireland. The end result may well be the Franco-Spanish force doing battle with the Channel Fleet off Dover and Nelson still cruising in the Mediterranean, but nevertheless we will have brought about a much more realistic, enjoyable and exciting replay of Trafalgar than can be gained by a mere laying down of models in imitation of the original deployments.

The explanation given above of the way in which we might refight Trafalgar well illustrates the advantages and disadvantages which naval wargaming has vis-á-vis land wargaming. When one considers the ease and spontaneity with which two 'landies' might set about refighting Waterloo it becomes apparent how much hard work and preparation is involved in playing naval wargames properly. The additional interest and excitement in fighting battles that one's own strategy has brought about and upon the results of which whole 'wars' depend ensure that the rewards are much greater. Once a player has tasted power in moving his fleets around the world and changing the destinies of nations he rarely wants to go back to his table-top armies.

Chapter 2

Equipment for naval wargaming

Before the newcomer to land wargaming can begin with his new-found hobby he must first buy hundreds of tiny figures, guns and vehicles and then he must paint them. This may well cost him a large amount of money and take months to complete. Sea wargaming is not nearly so demanding either in terms of cash or preparatory work but there is nevertheless still a minimum amount of equipment which must be assembled before we can begin. This equipment consists of:

  1 Model ships (naturally).
  2 A playing area (ie, the sea).
  3 Damage record cards.
  4 Measuring and turning devices.
  5 Dice.

Let us examine them in reverse order.

Dice

We use three different types of dice in our wargame; their different uses will become apparent later. The first type is the standard six-sided dice marked:

The next one is known as the average dice and gives a reduced likelihood of extreme results. It is a cube marked:

2	3	3	4	4	5

Finally we have the decimal dice. These are not cubes but decahedrons (ie, ten sides) or icosahedrons (20 sides) marked from 0 to 9. By throwing them in pairs and calling one dice the tens and the other the units we have the whole range from 1 to 100 (count 00 as 100).

Ordinary dice are generally available and although one will suffice it is best to provide a handful of them as they invariably disappear under the sideboard the moment they are required. Average and decimal dice are specially made for wargames and are available from most wargames or model soldier shops or by post from Navwar and from Skytrex, whose addresses are given at the back of the book.

Measuring and turning devices

This is simply a grandiloquent way of saying rulers and cardboard circles.

Ships have a minimum diameter of turn, which may vary with speed, with length and with a variety of other factors. In our sea battles we shall have ships trying to manoeuvre without colliding, attempting to ram the enemy and turning to avoid rams and torpedoes. If we are going to simulate battle accurately we must ensure that no model ship turns tighter than its real-life counterpart could do, so we use card turning circles marked around the circumference in inches or whatever unit of movement is used. In periods such as the two World Wars where long gun ranges prevent the ships coming into very close contact, one turning circle will suffice. But in, say, the Ironclad period where the ram is a much-used weapon and so the turning diameter of a ship becomes an important factor in its ability to attack with or avoid the ram, then circles must be provided in a number of different diameters to represent the turning diameters of ships of various lengths and suffering from varying degrees of damage. The circles can often be combined with measuring devices as shown below.

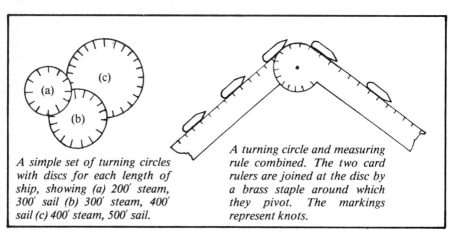

A simple set of turning circles with discs for each length of ship, showing (a) 200′ steam, 300′ sail (b) 300′ steam, 400′ sail (c) 400′ steam, 500′ sail.

A turning circle and measuring rule combined. The two card rulers are joined at the disc by a brass staple around which they pivot. The markings represent knots.

The straight-line measuring devices can simply be ordinary rulers or tape measures, but I find that moving the models a number of centimetres, or firing at a target at so many inches range takes away much of the realism of the game. It is only a matter of taste, but I much prefer to use my home-made devices, cardboard movement rulers marked in knots and dowel rod 'range finders' marked in hundreds or thousands of yards.

Damage record cards

The land wargamer has his battalions of model soldiers and as enemy fire causes casualties to them he keeps track of their declining strength by removing figures. With model ships, unfortunately, there is no way in which accumulating damage can be shown without recording it on paper. What form the damage record cards take depends on the complexity or simplicity of the rules used. In their simplest form they are just pieces of paper, perhaps torn from an exercise book, on which the ships' names are written; as damage points are inflicted they are written by the names and when a predetermined total is reached the ship sinks. Usually something more elaborate is required; perhaps a cutaway profile of each ship with engine rooms, magazines, guns, etc, shown so that they may be blacked out when they suffer damage. In such a case the damage record cards are too intricate and complicated to draw up at the beginning of a battle and throw away afterwards; some method has to be found which will allow the cards to be re-used. One simple way is to reproduce them by photo-copying but this can be expensive if large numbers are involved and unsuitable if colour is required. Probably the best method is to gum the original to cardboard, cover it with clear plastic and then a chinagraph pencil can be used to write on it and wiped off when finished.

A handy device known as a smoke marker is useful if used in conjunction with damage record cards. On the cards sufficient damage may be recorded to cause a loss of speed to some of the ships in the fleet. It would be a dreadful nuisance to have to look up the record card for each ship in the fleet before we could move it, so we put down a marker by those ships which are incapable of moving at full speed so that only their cards need be referred to. A counter or such like could be used, but a better visual effect is gained if the marker is made to look like a pall of smoke—cotton wool is very good for this purpose if teased out into the right shape.

The playing area

A land battle usually takes the form of two armies lining up facing each other, marching forward and fighting in the area between, so that the land wargamer, providing he restricts his armies to the size of battlefield, can play on any surface, small or large. He usually chooses a table top. Naval wargaming is not quite so simple in this respect; the sea battle frequently takes the form of a chase action or, if the battle is not a chase, Fleet Big-gun may well occupy itself in trying to open the range while Fleet Small-gun is busy trying to close it, which amounts to much the same thing. The result is that whereas the land battle is fought in one place the sea battle is often moving and so the biggest possible playing area will be needed and this effectively rules out a table.

The best place to fight a sea wargame then, is on a Mecca dance floor, but the cleared floor of one of the rooms in the house is more usual. If the room has a blue carpet the floor can be made suitably sea-like by laying down a large sheet of clear polythene and weighting it at the corners with books or furniture to stop it shifting. If your carpet is some other colour a sheet of blue polythene will look almost as good.

Problems of space can still arise on a living room floor, but these can usually be eliminated with a bit of forethought. The wind direction will often give a clue to the course a sailing battle will take and chase actions will usually head towards the fleeing force's base so that, by starting the action at one corner and having the diagonally opposite corner represent down wind or the way home, the best utilisation of space can be made.

Occasionally the pursuit will last too long or the battle change direction and the fleets will find themselves steaming full-ahead towards the skirting board. Under these circumstances I usually try to move the fleets relatively, so that if, say, an 11-knot squadron is being pursued by a 13-knot force, each 'game move' the slower force will remain stationary whilst its pursuers will be brought up by 2 knots per move, so that the range shortens by the proper amount but the end of the world, the skirting board, comes no nearer.

The method described above can be a bit clumsy, but I know of none better for use with largish-scale models. With very small models, though, such as 1:4,800-scale ships, a very good system employing hardboard squares to represent the sea can be used. In our rules for this scale 2 ft represents ten miles. Our campaign maps are gridded with ten-mile squares and on the floor the sea is represented by a number of squares of hardboard, with 2 ft sides and painted sea blue. Each sea square, then, corresponds with a grid square on the map. Where forces meet on the campaign map an appropriate number of sea squares are placed on the carpet and the model laid down on them in the same position that they lie on the map.

The battle is then fought out with models and as they steam along one square and look about to go over the edge, a sea square is taken from behind and placed in front of them. All that needs to be done to ensure a free-flowing and unrestricted battle is to keep a note of the grid reference which each sea square represents and carefully to slide the sea back across the carpet when the fleets approach the end of the room.

The sea squares provide useful reference points for plotting minelaying in a battle and noting the positions of submerged submarines. They also allow two contemporaneous actions, miles apart, to be fought simultaneously.

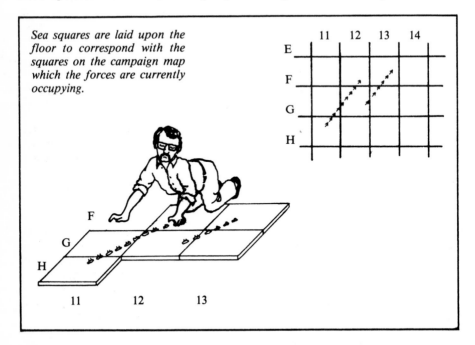

Sea squares are laid upon the floor to correspond with the squares on the campaign map which the forces are currently occupying.

Model ships

The most important thing to consider when buying or making models for naval wargaming is the scale to be used. As I have said before, the naval

A battleship gives fire support to a troop landing during the First World War. A correctly-scaled model would look completely out of place here, so the proportions have been distorted to make it look right.

wargamer is more likely to find himself embarrassed for space than the land wargamer and so we do not want to aggravate the problem by using extravagantly large models. At the same time, it is generally true that the smaller the scale of a model the less detailed it is and the less individually attractive and, although it really should not make any difference, most wargamers agree that the finely detailed, realistic-looking models always seem to fight better than the anonymous blobs. To coin a platitude, then, our models must be of the right size—no bigger and no smaller.

What is the right size? This will depend on three factors; the size of the playing area, the historic period of the game and the type of battle envisaged. The significance of the size of playing area is obvious—the wargamer with the use of a dance floor can use bigger models than the fellow restricted to a coffee table. The period is important because larger models can be used for the early periods when gun ranges are short and fleets prone to 'mix it' in a mêlée than in 20th century actions where the fleets stand off and fire away at ranges of ten miles or more. The type of wargame envisaged affects our choice because if we are buying or building cruisers for use in a *guerre de course* campaign where only single ships or small squadron actions are to be expected, then they can be much larger models than if they are expected to share the available sea with 20 or 30 other members of a fleet. Similarly, gunboats, monitors, landing craft, etc, built for use alongside model soldiers will have to be very large and of a distorted scale if they are going to blend in with the scene.

The accepted standard scales for model ships are 1:1,200 and 1:1,250 and the

HMS Hood *in four different scales—1:1,200 (Airfix), 1:2,400 (Starcast), 1:3,000 (Davco), 1:4,800 (Leicester-Micro).*

commercially available models seem more often than not to be of these sizes. For modern period war games though (by which I mean the First World War onwards) the models are far too big; a gun range of 20,000 yards at 1:1,200 converts to 50 feet and this is impossibly long even if we *do* have a ballroom to play in. The purists, in fact, often argue that, with such long ranges to scale down, models are best done away with altogether and pencil marks or coloured pins on a map substituted, but one thing I have never claimed to be is pure and so I continue to struggle along with Dreadnoughts in 1:3,000 scale, which is about the largest that can be conveniently used. Even so, much juggling with the speed and distance scales is necessary when using these models, and those in 1:4,800 scale would probably be better, although they are naturally less detailed.

The 1:1,200 scale can be used quite easily for ships of the Ironclad period where guns were unlikely to hit beyond a mile, and also for the 18th century sailing ship era, although the 'wooden walls' can also be modelled in a larger scale: photographs later in this book show my fleets of converted Airfix sailing ships which are in a scale of about 1:600. Ships of earlier periods can be modelled even larger still; although there are some excellent 1:1,200-scale ancient galleys available, which are usually used for a board type wargame. I prefer my ships of this period actually to carry model marines so my galleys are built to approximately 1:60 scale and, so long as the battles are of a limited size, this scale of models works quite well.

Until fairly recently the chap assembling his first wargame fleet was invariably compelled to make the ships himself as the range of suitably-sized models which were commercially available was very small. Nowadays the situation in this respect is much better. Special wargames model ships at low prices are now available for all the most popular periods, which usually means the First and Second World Wars, the 18th century and ancient times. A list, although not an exhaustive one, of the ranges available is given in the appendices.

If your interests are centred on the less well covered periods then you may still have to make your own models. Before the arrival on the market of the 1:1,200-scale Napoleonic ships most wargames of this period used conversions of Airfix's range of small kits of historic ships, and these are still of use. The kits seem to be of almost constant size rather than constant scale and the hulls are complete down to the keel, but for all that they represent a useful source of wargaming models. A hot knife is used to cut away the hull bottom and any other unwanted features, then use a file to clean up the rough edges that the hot knife left behind and the result is a waterline hull ready for painting. With lower masts halved and sleeved over with pieces of plastic drinking straws the masts and sails are made removable for displaying damage—a feature difficult to build into small models.

The *Victory* in this series has been used probably more than any other model in Nelsonic wargames, cut down to two decks, or transformed by lengthening into giant Spaniards. Also in the series are the *Revenge* and *Mayflower,* which are good for use in Armada period English fleets, and the *Golden Hind* and *Santa Maria* which, because of inconsistencies in scale, will pass for Spanish Great Ships and hulks of the period. By cutting down the fo'c's'les and poops and then rebuilding them with card and balsa the ships can be made to take on the towering individual appearances which distinguish the Spanish galleons of the era. One of the more ingenious of my old wargaming opponents actually hacked the Airfix *Victorys, Mayflowers* and *Revenges* about so much that he

The Airfix Revenge *kit. The hull and masts will be cut with a hot knife along the black lines.*

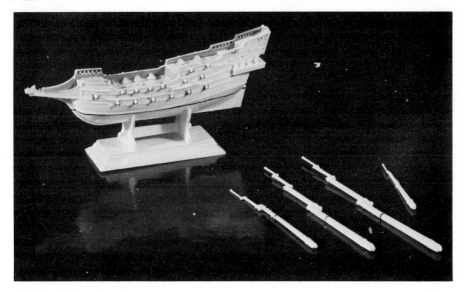

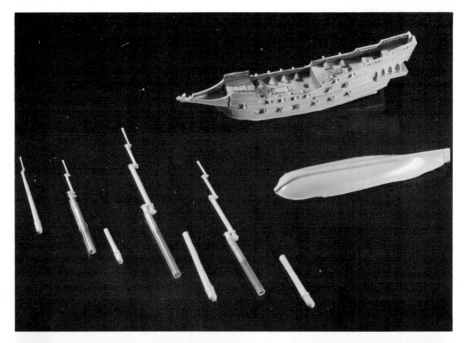

eventually produced, before an amazed and admiring group, two fleets of ships of the Dutch Wars. Hindsight makes me think, though, that the resemblance between the resultant models and the gilded battleships of the 17th century was largely in the eye of their fond creator.

When neither the possible conversions of the Airfix range, nor the products of the wargames equipment suppliers cover our period of interest, scratch building becomes the order of the day. This is not nearly as daunting as it may at first sound to the non-modeller as we are, after all, only aiming to make presentable, recognisable models rather than masterpieces of the miniaturist's art. Certainly the easiest and quickest way of building up a fleet is to make cardboard profile cut-outs of the ships required. The ships are drawn on to card and coloured in with crayons or felt-tip pens before cutting out with scissors and a craft knife and completing the other side; they are then affixed to a cardboard base so that they stand upright. They can look very good from the side but, of course, they almost disappear from sight when viewed from the end or above and this is their principal disadvantage. They are incredibly easy to make though; on one occasion a couple of years ago, when Paul Hanley and myself rashly decided to refight Tsu-Shima the following day, we made card cut-outs of all the Russian and Japanese battleships and cruisers engaged in the battle in the one evening—and we still had time to go out for a couple of pints before the pubs

Above Left *The hull has been reduced to a waterline model and drinking straw sleeves fixed to the upper halves of the masts.*

Below Left *The completed* Revenge.

Below Card cutout models of the Russian battleship Imperator Nicolai I *and the Japanese* Mikasa. *These models are very cheap to make and can look very realistic, but only from the side, of course.*

closed. The beer was paid for twice or thrice over by the money we saved as the two fleets together cannot have cost more than tenpence.

For all its frugal virtue the card cut-out fleet is unlikely to satisfy the critical eye for long. A much nicer model can be made from balsa wood by the bread and butter, or built-up, method of construction. This means that the shape of the ship is broken down into horizontal slices which are cut out of balsa sheets of appropriate thickness and glued one on top of the other.

Balsa sheet comes as thin as $\frac{1}{32}$ in so that in 1:1,200 scale the vertical dimension of a model can be accurate to within about 18 in of the original. This method of model making is best illustrated by demonstration so here is a step by step description of the making of a 1:1,200-scale model of the Italian Ironclad *Duilio* of 1878.

A more complex hull shape might have illustrated bread and butter construction better but I have been intending to build an Italian Ironclad fleet for some time and so *Duilio* it is going to be. She and her sister, the *Dandolo,* were the first of a number of strange and wonderful battleships built by the newly united Italy to counter the French threat. Unable to match the French in numbers the Italians opted for great fighting power in single hulls with the result that when *Duilio* and *Dandolo* were launched they were the fastest battleships in the world, with the biggest guns (4 × 17.7 in) and protected (where they *were* protected) by the thickest armour afloat. Just how powerful a fighting ship *Duilio* was, with much of her hull unarmoured and her monstrous muzzle-loaders only able to fire once in five minutes (and then singly for fear of damaging the hull) is questionable, but such floating white elephants add interest to any wargame fleet.

Modelling the *Duilio*

Research

I have in front of me *Jane's Fighting Ships* for 1906, *Man-of-War* by McIntyre & Bathe and *Warships—1860 to the Present Day* by H.T. Lenton. *Jane's* give us the dimensions of the ship and a photograph, *Man-of-War* gives us a plan view and profile in a large but unstated scale, and *Warships* gives us a nice, coloured profile drawing of the *Dandolo.* This is all we need.

By comparing the dimensions given in *Jane's* with the size of the plan and profile in *Man-of-War* we can calculate the scale in which they are drawn, 1:500. All we need do now to get our scale dimension is to divide those on the drawing by 2.4. The picture in *Warships* gives us the colour scheme for when we come to paint the completed model, it also gives us the detailed differences between the *Duilio* and *Dandolo* in case we want to build them both and, used in conjunction with the photograph in *Jane's,* it gives us an impression of what the ship looked like.

This 'feel' for the ship is very important when modelling in such a small scale: we are only poor amateurs when it comes to modelling and our proud creation is of necessity going to be a very much simplified model, with much of the minute detail left out. By looking at silhouettes and photographs we can see what it is on the *Duilio* which makes her look as she does and distinguishes her from any other ship but her sister. Fortunately for us she has a very distinctive appearance, principally characterised by a superstructure well aft, two enormous turrets amidships, her widely-spaced funnels connected by a light communication bridge, and a single military mast stepped amidships. She is covered all over, like most ships of this period, with ventilators but they do not contribute much to the distinctive look so that we can thankfully leave them off our model—they are almost impossible to reproduce. She also has light walk-ways forward, aft of the funnels which are not an important feature but which we will include because they are easy enough to add and because they will serve to distinguish our ship from her sister who does not seem to have had them. The boat cranes aft may or may not be important to the look, but we will include them just to be safe.

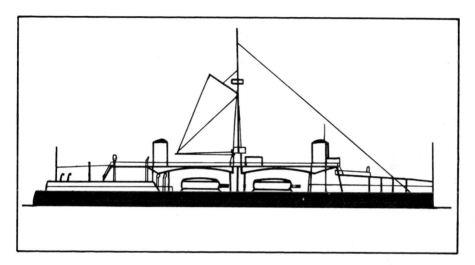

Materials and equipment

We need the following materials:

1 Balsa sheet $\frac{1}{8}$ in, $\frac{1}{16}$ in and $\frac{1}{32}$ in thick.
2 Pins, preferably of two types, ordinary straight pins, and shirt pins, which are thinner.
3 Thin card.
4 A plastic drinking straw 2–2½ mm in diameter.
5 Balsa cement.
6 Bostick and/or epoxy resin.

We need the following equipment:

1 A craft knife with a couple of sharp blades.
2 A pair of pliers with a wire-cutting edge.
3 A pair of tweezers.
4 Very fine sandpaper or flour paper.
5 A metric ruler.
6 Matt paints, brushes, etc.

Method

Cut the following pieces out of balsa wood and card as shown in the diagram overpage:

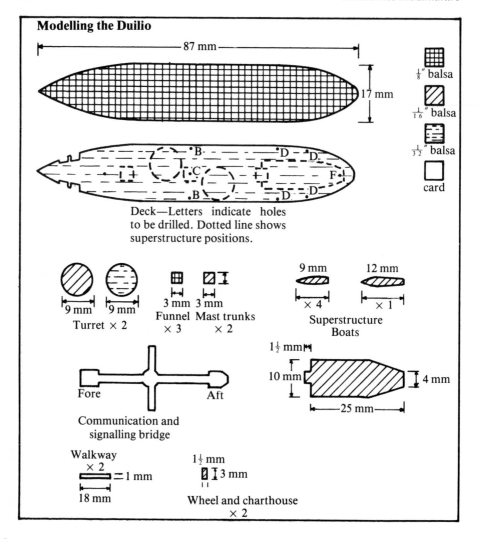

Modelling the Duilio

87 mm

17 mm

⅛″ balsa

1/16″ balsa

1/32″ balsa

card

Deck—Letters indicate holes to be drilled. Dotted line shows superstructure positions.

9 mm   9 mm
Turret × 2

3 mm 3 mm
Funnel Mast trunks
× 3   × 2

9 mm   12 mm

× 4   × 1

Superstructure
Boats

1½ mm

10 mm   4 mm

25 mm

Fore   Aft

Communication and signalling bridge

Walkway
× 2
= 1 mm
18 mm

1½ mm
3 mm

Wheel and charthouse
× 2

Glue the deck on to the hull, the $\frac{1}{16}$ in turret sections on to the $\frac{1}{32}$ in sections, the two $\frac{1}{16}$ in funnel/mast trunks on to two ⅛ in trunks and the remaining ⅛ in trunk on to the forward of the superstructure. Sand and smooth.

Glue the superstructure and forward funnel and mast trunks into the position shown on the deck diagram but not, at this stage, the turrets.

Drill the holes in the deck and superstructure as shown, and also two holes, approximately 3.5 mm apart, in the face of each turret to take the guns. This drilling is quite easily done with a thin, heated pin; the holes are neatly burned into the wood and this prevents the balsa splitting when the pin is inserted.

Insert thin shirt pins to represent bridge and walk-way supports in holes A, B and E, and affix with epoxy or Bostik. Supports A and E should stand 2 mm high, supports B $\frac{3}{16}$ in, or the height of the funnel and mast trunks.

Glue the communication/signalling bridge in place on top of the funnel mast trunks and supports B, using epoxy or Bostik. Ensure that the square end is to the front and overlaps the forward funnel trunk by about $1\frac{1}{2}$ mm and that the centre of the 'cross' of the bridge is over the hole drilled in the mast trunk. Glue the forward walk-way underneath the forward end of the bridge and to the top of pin A. Do the same with the after walk-way and pin E.

Cut two 4 mm lengths of the drinking straw and fix with epoxy to the bridge on top of the fore and aft funnel trunks. Glue the wheel-house on the bridge immediately in front of the fore funnel and the chart-house on the bridge just ahead of the mast trunk.

Glue an ordinary straight pin into the hole in the mast trunk and cut short 12 mm above the top of the bridge. Cut the head and point off a thin shirt pin so that it is 15 mm long, glue it with epoxy resin to the top of the thicker lower mast pin so that they overlap 5 mm with the upper pin to the front. A small circular fighting top, approximately 3.5 mm in diameter, made out of scrap plastic is split in two, the centre hollowed out and then glued around the join to strengthen it and support it whilst it sets, which will take around 24 hours.

Bend four ordinary pins to shape as boat cranes and fix into place with epoxy in holes D. Stick a shirt pin in hole F and cut to 8 mm high—the ensign will be attached to this staff. Fix ordinary pins into the turrets as gun barrels and cut to 2 mm length.

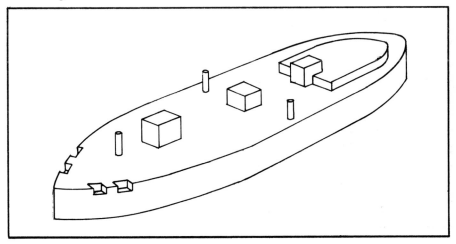

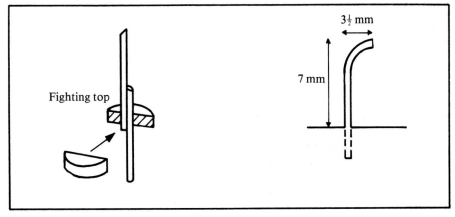

The completed model of the Duilio. *The smoke is made from pipe-cleaners and cotton wool blackened with ink.*

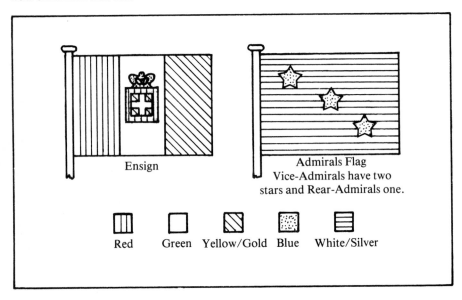

Ensign

Admirals Flag
Vice-Admirals have two
stars and Rear-Admirals one.

Red　　Green　Yellow/Gold　Blue　White/Silver

With the turrets and boats still separate the ship is painted to the following colour scheme:

Hull sides, bridge supports, cranes, barrels—black
All decking, bridge, superstructure deck—creamy white
Funnels, funnel and mast trunks, mast—buff
Superstructure sides, boats, turrets—white
Wheel- and chart-houses—brown with silver windows

Also paint portholes in the superstructure side and thwarts on the boats.

When the paint is dry, glue the turrets and boats into place. Cut small flags out of notepaper, a rather oversize 3 mm × 4.5 mm allows easier painting. Glue one with Bostik to the staff aft to be painted as an ensign and one, if it is to be a Flagship, to the mast-head.

Chapter 3

Rules for naval wargaming

Once upon a time, in the early days of popular wargaming, it was almost inevitable that a player, or at least a group of players, would write their own rules for the particular period in which they were fighting. These home-made rules were often quite simple, and occasionally naïve, but they did represent the wargamer's interpretation of the warfare of his period, and they were well tailored to the type of game which he played.

In recent years it has become increasingly usual for wargamers to go out and buy their sets of rules from one of the many suppliers of wargames requisites which the growth of the hobby has thrown up. These commercially produced rules have many advantages, particularly for the wargamer whose activities are linked to a wargames club. They are invariably well thought out by people who know their period inside out; they are easily obtained and generally accepted so that the wargamer playing 'away' and his 'at home' opponent or fellow club member can agree on the rules to be used and familiarise themselves with them days before the battle; and they are usually extremely comprehensive so that in tournament games and other needle matches where the spirit of disinterested concession wears thin, disagreements are minimised by having almost every circumstance legislated for. The unfortunate consequence of the search for complete accuracy and comprehensiveness, though, is that the rules become bulky with pages of exhaustive regulations, and in the wargame more time is spent thumbing through charts and consulting tables than is spent in manoeuvring the models.

To the wargamer who fights at home against one, or perhaps two, regular opponents standardisation and exhaustive completeness of rules are not too important. Regular opponents who battle against each other weekly and then discuss their triumphs and disasters afterwards over a pint are usually able to bring a spirit of friendly compromise, larded with a knowledge of the historical period, to bear on those novel situations which arise and which are not adequately catered for in the rules.

The attempts at super-accuracy will also often be found to be counter-productive. The need to be constantly searching through tables to bring into account each and every factor that affected real-life combat will cause the game to progress so slowly that tactics become forgotten for fear of extending the tedium. The essence of realism in a wargame is not the inclusion of all the details, major and minor, which had a bearing on the course of battle, but the sensation of decision making, of events flying along too quickly with too little time available to think them out and guide them. Our rules should be geared to achieving this kind of wargame, even at the expense of accuracy, otherwise our wargames will become as formalised and unlike war as chess.

To the non-club wargamer then, I would suggest writing one's own rules, designed for fast-moving actions without a too strict regard for the detailed features of the period. The drawing up of a suitable set though, is not always easy. The characteristics which I always aim for when writing rules (though they are rarely achieved completely) are that the two main parts of the game move, moving and firing, should be made sufficiently simple and straightforward to allow of their being carried in the players' heads, so that no time-wasting rule-book checking is necessary. Those minor areas which are referred to less frequently, say, those relating to morale, or coaling, or mine-laying, whilst ideally being easily memorised, may be more complicated and can be looked up in the rule-book.

The trouble is that an *over*-simplified set of rules often produces wargames which display little of the flavour of the period which they are meant to represent. We once simplified our firing rules for First World War actions to a system whereby each ship had a fire factor of which, varying with a dice throw and range, a proportion was deducted from the target ship's floatation factor. The system was admirably simple, quick and easy to use; the results obtained were just as accurate as any other method we have used. Nevertheless after a couple of games we reverted to the earlier, much slower, firing table system. The new system had made no pretence to be simulating salvo firing and so it had not seemed to our minds that the ships were firing at all—just emanating a malign influence over a distant enemy which caused them to weaken and sink.

If really big battles are going to be fought, though, it is essential to go some considerable way to synthesising the rules in this fashion as otherwise the sheer fag of having to fire the guns of a large fleet will cause the Admirals to commit the most stupid tactical blunders through pre-occupation with rule checking. The games will be boring too.

On the other hand certain types of wargame do require very 'analytical' rules; small actions, for example, with only one or two ships a side. In these battles there is little scope for the application of grand tactics and so, to make the wargame enjoyable and skilful, the characteristics of the individual ships have to be strongly highlighted so that the Captains may properly exploit the strengths of their own ships and weaknesses of the enemy's to secure a victory. Otherwise such actions would be just uninteresting slogging matches with the strongest ship, ie, that with the largest 'fire factor', winning regardless of the skills and competence of the player.

Any campaign, or even a series of unrelated games, in a given period will inevitably produce a number of battles ranging from single ship duels to gigantic fleet actions; how is one to decide the level of 'analysis' or 'synthesis' to build into the rules? The solution we have adopted is to use two or three, or more sets of rules for the same period and employ that set most appropriate to the size and type of battle. In this way we can fight great dynasty-shaking fleet actions and small, insignificant cruiser and frigate duels with equal ease and a similar amount of interest.

Our basic approach to devising sets of rules for a given period is to ask ourselves what person will the player in this type of wargaming represent, and then we decide which decisions that man would make himself and which he would leave to subordinates. If drawing up rules for First World War fleet actions, for example, we would cast the players in the roles of Commanders in Chief of large battlefleets whose responsibilities would be: directing their cruisers to find the enemy, deploying and manoeuvring the battle squadrons to gain the advantages of position, smoke, light and wind, and giving general instructions to their cruiser and flotilla commanders for detailed interpretation by themselves. Our rules therefore, would concentrate on signalling methods and the effects of light, smoke and wind on gunnery; the actual firing of guns would be a relatively simple affair and such matters as damage control, which would in reality be left entirely to subordinate officers, would be treated as an automatic operation.

If, on the other hand, we were drawing up a set of rules for single-ship actions in, say, the Napoleonic era our approach would be quite different. In such wargames the player represents not an Admiral with a host of subordinate officers to relieve him of much of the thinking, but a Captain with a single ship to look after. These rules would largely ignore reconnaissance, signalling and so on but would concentrate on the principal areas of the Captain's duty, manning the various parts of the ship for optimum performance, sail control, fighting tops, port and starboard broadsides, damage control and boarding parties. They would also deal in a detailed manner with the performance and handling of the ship under sail. The sum effect of this practice of ours is to give the wargamer a roughly similar amount of work and decision making in each game which arises, regardless of its size.

A good way of bringing additional fun into one-ship-per-player games is to make the actual firing of the guns a business of individual skill on the part of the players rather than the usual weighted-chance system. Probably the best known of the 'all-skill' methods is that of guessing the range. This system in its simplest form means that each player, standing behind his ship, guesses the distance to the target vessel and the better his estimate the more shells of the salvo hit. Misses are denoted by the umpire putting down splash markers for the shooter to correct from. This method is best used on a very large playing area (we use a lawn in summer) because range finding becomes too easy in a small space, as it does if the players are handling tape measures, so that ideally an umpire should measure the ranges.

We soon found ourselves getting too good at this method so we complicated it by bringing deflection into it. Given the very long ranges in naval gunnery in this century it could take up to a minute for a batch of shells to reach the target, so that the art of firing was not simply to calculate the range properly but also to predict where the target would be when the shells reached it.

We simulated this by laying out a firing arrow (pegging it to the lawn) and writing the estimated range upon it at the beginning of the move, then moving the models, and *then* measuring the angle and range. The line indicated by the firing arrow was followed exactly and the range was measured from it, ie, where the firing ship was at the beginning of the move. The rules required changes of speed or course to be noted two moves ahead, and turns of more than $22\frac{1}{2}$ degrees prevented guns firing, so as to discourage too vigorous zig-zagging. Misses were indicated with splash markers and the number of hits was calculated with a perspex disc marked with radiating circles indicating the number of hits, more towards the centre, fewer towards the edge.

Another all-skill method which we have used with success was devised for use

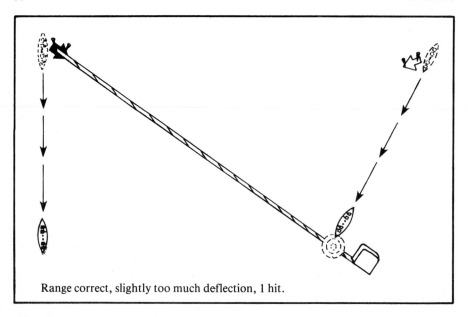

Range correct, slightly too much deflection, 1 hit.

in a game in which a number of the members of a wargames club participated. Four players a side each took command of an Ironclad battleship. One player on each side was elected Admiral and the only method by which he and his Captains were allowed to communicate was by chalking on a board the number of a signal written down before the battle in a signal book.

The firing was carried out more or less according to the rules given in Chapter 8 but it was done by throwing darts at detailed profile drawings of the ships pinned to sheets of plasterboard, the darts being thrown from between 3 and 12 ft depending upon the range measured between the models. Raking fire was carried out by requiring the firer to throw his darts at an angle to the drawing; fleeting, passing shots were simulated by having the player spin round and throw all his darts in five or ten seconds, whilst chucking the darts with the left hand was thought a good way of representing fire from a steeply heeling, sinking ship. Darts may not represent a very life-like gunfire effect but the system does allow a group of people all to participate enjoyably in a wargame, without the problems which usually occur when they take part in a conventional game.

Chapter 4

Ancient galley warfare

For about 3,000 years, up until the mid-18th century, the oared galley was the principal warship of the Mediterranean navies, but it is the period 500 BC to 100 BC, the heyday of the ram, which is principally catered for in this chapter. Before this period warships were regarded less as missiles in themselves than as platforms for carrying soldiers into boarding action with other soldiers and after 100 BC the size and weight of galleys started to grow to the extent that speed and manoeuvreability were impaired and the value of the ram declined accordingly.

The war galleys of our period, the age of the Persian and Punic wars, were lightly constructed vessels of about 100 to 150 ft in length and were propelled by oars in two or three banks (biremes and triremes respectively). A large square sail could be hoisted to assist the oarsmen, but this was only of use when the course of the ship and the wind direction more or less coincided, because with the clumsy sail and shallow draft the galleys were quite incapable of sailing into the wind. Indeed it was common practice to lower the mast when rowing into a headwind to reduce resistance. With a fresh crew pulling hard a speed of about 7 knots was possible, but this would fall off very quickly as the crew tired; a 2- or 3-knot cruising speed would be more usual.

With over 200 men, oarsmen, sailors and marines, in what was basically a big open boat, it was necessary for the crews to go ashore at night to cook and sleep. This close dependence on the land, however, was not so serious an inhibition to strategic movement as it might seem because contemporary navigation was so crude that journeys could only be made by taking short hops from island to island and headland to headland, and the ships themselves were too light and low to be considered as anything but inshore craft. This light construction did have its benefits as docks were completely unnecessary to even the largest trireme as they could and would be hauled up stern first on to a beach at night and relaunched the following day.

In addition to the rowing crews who, contrary to popular belief, were usually paid volunteers, there were soldiers aboard whose numbers tended to vary with the importance which they were accorded in the fighting philosophies of their fleet. In the navies of the Phoenicians, of Carthage and of many of the Greek states, where there were large and skilful seagoing populations, the ram was regarded as the great battle winner and boarding as being very much a secondary tactic, so that the number of marines was kept quite small. In contrast, those landlubbers the Romans found themselves unable to emulate the subtle and sophisticated tactics of other fleets and so attempted to take land warfare to sea by packing their ships with marines and fighting by laying their ships alongside the enemy and overwhelming him in a boarding fight. In the war against Carthage the Romans even went so far as to equip all their ships with spiked drawbridges, known as *corvii*. When an enemy came close enough the *corvus* was dropped on him, the spike dug into his deck and held him and the marines would pour across the gangway and take the ship. The Romans also constructed towers on their ships to allow bowmen to command the enemy's decks with their archery during ramming or boarding attacks; it was only towards the end of our period, however, that the introduction of stone- and bolt-throwing engines aboard ships allowed stand-off battles to be fought.

When the fleets deployed for battle they were formed up in line abreast. This formation had two purposes, it ensured that all the ships' rams were facing the enemy and it meant that the sides of the galleys, with their rows of vulnerable oars, were covered by their neighbours in the line. The space allocated to each ship in the line was dictated by the need to allow each galley enough room to turn independently whilst keeping them sufficiently close together to give protection to oars. Consequently advancing through a narrowing channel in battle formation could cause chaos. Forming up as they did in line abreast, and fighting to the front rather than to the side as in later period ships, and having the abilities to stop, advance and retire, the galley fleets in many ways bear closer resemblance in their fighting methods to armies than to navies of the gunpowder age. As the distinctions between sailor and soldier, and Admiral and General, in those days were non-existent it is not surprising, therefore, that sea tactics were often similar to land tactics, with channels instead of defensible defiles and shore lines being used to rest a flank upon where an army would use a river or wood.

In the early days of galley fighting, battles took the form of the two fleets facing each other in line and then simply rowing into a head-on mêlée. By the time of the Peloponnesian War, however, these unimaginative tactics had developed, with the objective of attacking the enemy in his vulnerable rear, into the *diekplous* and the *periplous*.

The *diekplous* required considerable co-ordination plus an element of surprise. On a given signal one of the galleys in the centre of the line would race forward at the space between two of the enemy opposite. Putting her helm over, the prow of the galley would scrape down the side of one of the enemy, smashing all her oars on that side and effectively crippling her so that the galley following the first attacker could easily ram and sink her. Through the gap thus created would follow all the ships of the attacking fleet, turning left and right alternately and attacking the now thoroughly disorganised enemy in the rear.

The *periplous* was a tactic particularly suitable to a fleet in superior numbers as it required that the flank of the enemy fleet should be overlapped by the flank of the attacker. Whilst the galleys facing the enemy would pin him, those of the overlap would turn the enemy's flank and attack him in the rear.

The great strength of the *diekplous* and *periplous* was that by taking precautions against one form of attack an enemy laid himself open to the other. The natural defence against the *diekplous* was to form a second line which would attack galleys emerging from the breach in the first line, but this, of course, reduced the frontage of the fleet so that it became vulnerable to a *periplous* attack. In the absence of shore lines or islands to rest flanks upon, the natural way of preventing a *periplous* attack was to lengthen the line by increasing the spacing between ships but this only made the *diekplous* more likely to succeed. A formation to counter both methods of attack was the *kyklos,* the naval equivalent of the square, which Themistocles successfully used at the Battle of Artemesium; this involved the fleet forming up in a circle with sterns pointing towards the centre, and bows outwards. The *kyklos* was only defensive, of course.

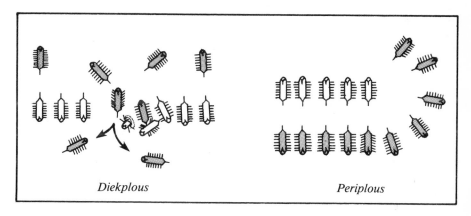

Diekplous   *Periplous*

The rules

The navies of ancient times were coastal fleets and tended to act as adjuncts of, and in close co-operation with, the armies ashore and their activities reflected this limited role, rather than the wider ranging independent strategic operations of fleets in more recent ages. My ancient naval wargames follow this pattern and are always conducted in conjunction with land campaigns involving armies of model soldiers. It frequently happens in these campaigns that the circumstances of a battle require that both naval and army units appear together in the resultant wargame and so my rules for fighting galley battles are designed, not around large fleets of tiny 1:1,200 models operating over a gridded 'sea' as is usual today, but around large models actually carrying seamen and marines in the form of 15 mm model soldiers. The rules can easily be adapted for use with the very tiny models, if this scale is preferred, with the actual fighting qualities of the crew reduced to factors.

In order that the rules may be easily understood it is necessary to know a little about the models for which they are designed. My galleys are made of balsa wood and card, and are about 6 in in length and $1\frac{1}{2}$ in across the beam. For obvious reasons oarsmen are not included in the ships, but removable oar sections, five a side, made of pins and card, slot into the black-painted wells where the rowers would sit. These sections can be taken out to simulate damage, as can the rudders hooked over the gunwales at the stern. Masts, made out of plastic drinking straws, and cardboard sails, can be erected (fitted over a stump amidships) or removed and optional fighting towers can be placed aboard the ships. The crews usually consist of a Trierarch (Captain), two seamen, two, four or six heavy infantrymen and a couple of archers.

Galleys assembled and dismantled. Masts and sails can be raised and lowered, oar sections removed to show damage and crewmen taken off to denote casualties. The fighting tower can be mounted or left off as required.

The sequence of play

The game is played more realistically and faster if, instead of the players moving their fleets alternately, they move them simultaneously. It is not inconceivable that an unscrupulous player might hang back and watch what the other player is doing before committing himself; it is best not to have such a person as an opponent, but to remove any temptation to indulge in such a dubious practice we have each player write down his orders to each of his ships and then declare them before moving. The sequence of play, then, is as follows.

1 Write down orders for each ship.
2 Declare orders and then move ships.
3 Archery and the firing of catapults takes place and casualties are removed. Firing is regarded as taking place simultaneously so that any archers shot in a move may reply before being removed.
4 Calculate ramming damage.
5 Boarding and mêlées.
6 Holes stopped, oarsmen transferred, etc.

The writing down of orders sounds a tedious business but is not necessarily so. All it is intended to do is to commit the ships to certain actions, once those actions have been decided upon, so that a shorthand system of symbols could be used. Here are a few that we have evolved over the years:

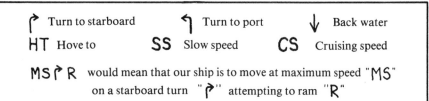

BATTLE *Ostia* FLEET *Arkadian* ORDER SHEET									
NAME	ENERGY	DAMAGE	1	2	3	4	5	6	7
Sum	ɣ⅃Ɉɣ⅄6ȿ		ss↑	ss↑	ss↑	ss↑	cs↑	cs↑	cs↑
Es	ɣⱮɈⱮ9		ss↑	ss↑	ss↑	ss↑	ss↑	cs↑	Fs↑
Est	ɣⱮ II		ss↑	ss↑	ss↑	(ⁱss	↑ss	ss↑	cs↑
Sumus		ⱵⱵ	ss↑	ss↑	ss↑	ss↑	SUNK		

These orders are written on pieces of paper, with a horizontal line for each ship and vertical columns for each movement period. We call this an order sheet which is a convenient title, but not a wholly correct one, as you will see on the example above that not only do we record orders, but also energy and damage points. These two are explained later.

Movement

Sailing

The ancient war galley was not a sailing ship in the true sense of the word and in battle its sailing gear would often have been left ashore, or at least have been stowed away. Nevertheless sails can be useful in an action for conserving oarsmen's strength or allowing crippled ships to break off action, so I include a simple set of sailing rules here.

In a campaign battle the wind direction will already be known but in a one-off battle it may be necessary to select it at random. Nominate one part of the room as North and the throw an ordinary dice—6 is a north wind, 5 east, 4 south, 3 west, 2 is either north-west or south-west, 1 is either north-east or south-east. If a 1 or 2 has been thrown, throw the dice again, 4, 5, 6 is north-west or -east, 1, 2, 3 is south-west or -east.

Sea warfare enjoyed a close season during the winter months and the ships rarely ventured so far from land that they could not race to the shore before bad weather engulfed them, so we will ignore the possibilities of very turbulent conditions and simply say that when a dice is thrown for wind strength 1 indicates a calm (throw again after five moves), 2, 3 or 4 indicate a light breeze and 5 or 6 a breeze. With the wind behind a galley and within 45 degrees of the centre line it may be moved 18 in by a breeze and 12 in by a light breeze. With the wind between 45 and 90 degrees of the beam, the galley may be moved 12 in by a breeze and 8 in by a light breeze. Under no conditions may a galley go closer than 90 degrees to the wind with its sail up.

The actual speed of the ship under sail will also be affected by damage to the hull, but this is dealt with under 'ramming'.

The sail may be raised or lowered in two moves and the mast in an additional one, by two men, one of whom must be a sailor. If no sailor is available to take charge of the operation two soldiers may muddle through in three moves and two moves respectively. The sail has no effect until it is fully hoisted and its effect does not cease until it is completely lowered and stowed away.

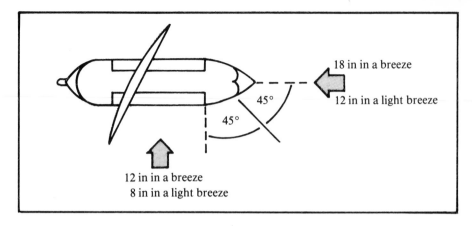

18 in in a breeze

12 in in a light breeze

45°

45°

12 in in a breeze
8 in in a light breeze

Movement under oars

Although it was usual for oarsmen aboard galleys in pre-Roman times to be freemen we will suppress our liberal consciences in this instance and declare all our rowing crews to be abject slaves, shackled and manacled. This means that the opinions and loyalties of the oarsmen can be conveniently ignored when prizes are taken and our boarding actions are saved the complication of rowers springing to the defence of the ship.

Energy

Unlike the Admirals of later ages who covered their distances as the wind allowed (or with the unflagging stamina of steam), the fleet commanders of classical times had to give much prudent thought to their movements in the days prior to battle, so that their galleys would arrive at the 'battlefield' with their oarsmen fresh and able to give of their best. This husbanding of rowing power is of great importance in a campaign and so we need a system of determining whether the pre-battle manoeuvres have left the oarsmen fresh or tired. What we do is to count up the number of hours rowing that the galleys have done in the five days before the battle and take note of the speeds at which they were rowing (speeds are explained later).

For each hour spent rowing count up the following: at slow speed—1 point; at cruising speed—3 points; at fast speed—7 points.

Total them for the previous five days:

> Up to 30 points the crew is fresh.
> 31 to 40 points the crew is tired.
> 41 to 50 points the crew is fatigued.
> 51 to 55 points the crew is exhausted.
> 56 points or more and they have died of exhaustion!

When the battle is joined each galley is allocated a number of energy points according to the freshness or otherwise of the oarsmen. (A column for recording these energy points is shown on the order sheet shown on page 35). Energy points are deducted for every move in which the galley moves at more than its slow pace, minus 1 for every move at cruising speed and 2 for every move at fast speed. When the crews' energy is exhausted all they then have the strength to do is plod along at slow speed.

Energy points are allocated as follows:

Fresh crew	—	12 points.	Fatigued crew	—	4 points.
Tired crew	—	8 points.	Exhausted crew	—	Nil points.

Should a galley, inadvertently or otherwise, be ordered in a move to go at a faster rate than its remaining energy points will allow, it loses one of its ten oar sections for every point that it is in debt. We call this 'dying of exhaustion' which is reasonable enough, even though it is designed to discourage inadvertence.

Speed

Speed may only change by one 'speed band' per move, except, of course, when the galley is brought to a sudden and violent halt by ramming another. The speed bands are:

> Backing water (at slow speed).
> Hove-to (ie, stationary).
> Slow speed.
> Cruising speed (costs one energy point).
> Fast speed (costs two energy points).

Thus if a vessel has just backed water, perhaps to withdraw its ram from a foe, and now wishes to move forward at cruising speed the sequence will be: next move stationary, second move slow speed forward, third move cruising speed forward. This same procedure is followed in reducing speed.

The maximum distance which a galley can move in each speed band will be reduced as oar sections are lost through archery fire, oar raking or exhaustion. The number of oar sections to count, if the complement is not complete, is the number on the side with the least, so that if, say, the galley *Sumus* has five oar sections to starboard but has lost two on the port side it will move as three until its oars have been redistributed, when it will move as four.

Speed table

	Oar Sections				
Speed	**5**	**4**	**3**	**2**	**1**
Slow and Back water	9 in	6 in	3 in	—	—
Cruise	15 in	9 in	6 in	3 in	—
Fast	24 in	15 in	9 in	6 in	3 in

The distances given in the table are only the maximums allowed, so that a fully-oared galley moving $9\frac{1}{4}$ in is regarded as being at cruising speed, even though by only a quarter of an inch, and so will have to use up an energy point.

Turning

Turning is one of those areas in which we can, if we wish, distinguish between skilled and unskilled seamen because the ability to out-turn one's opponent gives great advantages in a ramming battle. There is an historic justification for this rule in that the Carthaginians at least were able thoroughly to out-manoeuvre the Romans on occasions by using the rudders of both sides of their galleys simultaneously; the less skilful Romans were unable to co-ordinate the efforts of two helmsmen. If, therefore, one of your fleets is more amateurish than the other, make them turn on arcs one or two grades larger than those specified here.

A	—	4 in diameter	D	—	10 in diameter
B	—	6 in diameter	E	—	12 in diameter
C	—	8 in diameter	F	—	14 in diameter

Galleys moving at slow speed, or backing water, are assumed to be co-ordinating the oars with the rudder and so may turn on circle A (or larger if preferred). At cruising speed the tightest turn is on circle B and at fast speed the tightest turn is on circle C. All this assumes that a sailor is at the helm, but if for any reason a soldier has taken over the turn will be one grade larger, and if both rudders are destroyed and turning is by oars alone a circle two grades larger will be used. A vessel under sail will turn on circle D.

Phased moves

When one galley is being moved up to ram or board another, it is more than likely that the intended victim will want to avoid contact. The normal move in which the whole distance is moved in one bound does not properly reflect the way in which two ships in such circumstances would be moving and reacting to the enemy's movement, and so under these circumstances we use the phased move. All this means is that instead of moving those two vessels their full distance at one go they are moved alternately at the rate of a third, or a sixth, or a quarter of their move until contact is made or the whole move used up.

Fighting

Using the ram

Oar raking—Oar raking takes place when one galley attacks another at a shallow angle so that the attackers' bow scrapes along the other ship's side, smashing its oars. When the two ships come into contact they are moved, by phases, the whole distance of the move. The attacker smashes all those oar sections with which his bows have come into contact within $\frac{1}{2}$ in of their rowlocks. More than $\frac{1}{2}$ in from the rowlocks and the oars are assumed to give without breaking.

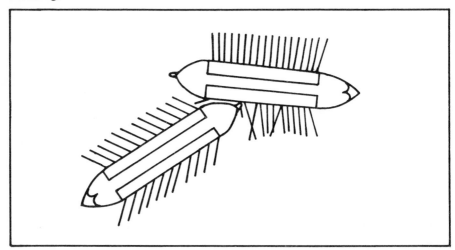

Ramming—Ramming attacks should be made at a more nearly perpendicular angle to the victim's broadside to be successful. The attacker and victim are phase moved until the ram strikes. If the ram hits amongst the oars, one section is destroyed, whatever other damage is done, and if it hits abaft the oars the rudder on that side is destroyed.

The angle of attack and the speed of the rammer are taken note of and then two average dice thrown:

Add 1 if the rammer is moving at fast speed.

Subtract 1 if the rammer is moving at slow speed.

Subtract 1 if the attack is made between 60 degrees and 30 degrees from the perpendicular (angle B).

Subtract 2 if the attack is made between 30 degrees and 0 degrees (angle C).

Angle A is where the ramming ship is perpendicular to its target.

Damage to the victim is as follows:

Dice total 11—the victim sinks instantly.

Dice total 10—victim holed and ships four loads of water.

Dice total 9—victim holed and ships three loads of water.

Dice total 8—target holed and ships two loads of water.

Dice total 6 and 7—target holed and ships one load of water.

Dice total 5 and less—target not holed.

When a ship has been rammed and holed it will take in one load of water per move until it has a total of five swilling about inside it at which point it will sink. The crew can make an attempt to plug the hole with a sail or something else equally suitable—throw an ordinary dice, at the end of each move during which the plugging is being attempted; if a seaman has been allocated to the job a 5 or 6 indicates that the plugging is successful, if a soldier is doing the job a 6 means success. Less than 6, or 5,6, means that water continues to flood in, try again next move (if you are still afloat!).

A holed and waterlogged ship will lose a quarter of its move for every load of water that it has aboard, whether under oars or sail, so that a galley which has plugged its hole in the nick of time but has four loads of water in it will remain dead in the water.

Supporting sinking ships—Two ships, one on either side, grappled to a sinking ship can support it indefinitely. One sound ship grappled to one that is holed will give support until the damaged ship has taken seven loads of water aboard at which point it will go under and drag the other one with it, unless the grappling lines are cut in time.

Abandon ship—Although a galley is sinking there is still a chance of saving the crew (oarsmen go down chained to their benches). For each man in the water throw one ordinary dice to see what happens to him—4,5,6 they can swim, 1,2,3 they drown. Swimmers can move at 2 in per move for 12 moves during which time they can save themselves by reaching shore or reaching a friendly ship, which must be hove-to in order to pick them up. Swimmers are naturally assumed to have abandoned their weapons and armour and so may not fight after being rescued, but they may, after recovering for two moves, assist in steering, plugging holes, handling the sails and so on.

Boarding

Movement aboard ship—Men may move anywhere on the ship in one move but are assumed to reach their final position, if it is more than about 2 in away, at the end of the move, so that a seaman moving the length of the ship to take over a suddenly vacated helm will not be able to begin steering until next move.

Grappling—Grappling lines are represented by lengths of cotton. When any ship comes within 2 in of another, any man (except Captains and Admirals) not otherwise engaged may throw a grappling line a maximum distance of 2 in. For each line throw a dice—3,4,5,6 mean that the hook has caught, 1,2 that it has not held. Two lines are needed to hold a ship; if only one line connects the ships it will snap if one of them rows off. Grappled ships will draw together and boarding may take place.

In subsequent moves unengaged men aboard the grappled ship may attempt to cut the lines—4,5,6 on the dice means they are successful, 1,2,3 try again next move. At this stage all lines must be cut to free the ship.

A *corvus,* when dropped on to an enemy's deck automatically locks the two ships together.

Boarding—When two ships are grappled together it is usually fairly clear which side is boarding and which side defending, but in a case where both sides are claiming the right to board the decision is carried out this way. If one side has noted on his order sheet the intention to board then he is the boarder, or if one ship belongs to a fleet operating the 'land battle at sea' principle of fighting then he is the boarder if his opponent is of the ramming school. If neither of these cases applies, or if they apply to both, the decision is made by each player throwing a dice and the highest scorer winning.

When boarding takes place Captains, heavy infantrymen, archers and even Admirals may take part, but seamen must remain on their own ship unless the intention is to abandon it. Where the point of contact between the two ships is blocked by defenders the attackers must fight their way aboard, stepping into the places of dead men, as described below. If the crossing place is undefended the boarders may just cross over.

Hand to hand fighting—All hand to hand fighting is carried out, as far as possible, as a number of duels between men of opposing sides. Naturally the numbers taking part, or the circumstances of the fight, will not always allow opponents to be paired off so neatly, so where necessary two men may fight one but greater odds than this are not allowed; a third man, it is assumed, simply gets in the way of the other two. It should always be ensured, in two-to-one fights, that the second man can actually get to grips with the opponent; there will always be circumstances, say when defending the top of a ladder or the central gangway of the ship, when the narrowness of the approach prevents more than one attacker getting to him at a time. In marginal cases our sense of fair play usually leads to a decision in favour of the outnumbered hero.

A 'normal' duel is one between two men equipped in the same way and of the same rank. (This latter is not strictly necessary, but it offends our sense of social order to have grand folk like Captains and Admirals fighting on equal terms with humbler sorts.) A normal duel then, would be a heavy infantryman versus a heavy infantryman, an archer versus an archer, a seaman versus a seaman, a Captain versus a Captain, or an Admiral versus an Admiral. In such cases throw one dice for each man—if one side gets a two-point advantage (eg, 5 and 3) or greater, he kills his opponent, a one-point advantage or less means that the fight is drawn for this round.

When a duel takes place between two men of unequal equipment or rank, the superior man adds 1 to his dice, but he only adds 1 however many degrees of superiority he has over his opponent. For example, an archer fighting a seaman will add 1 to his dice, a heavy infantryman will add 1 to his dice if he is fighting an archer and a Captain will add 1 to his dice if he is fighting a heavy infantryman, but so will an Admiral only add 1 to his dice if he condescends to fight a seaman.

Where a fight consists of one man fighting two the rules are a) 2:1 are the maximum odds, and b) 1 is the maximum which may be added to a combat dice. Otherwise, the man who is helped by a friend will add 1 to his dice. So, two heavy infantrymen attack one heavy infantryman and add 1 to their dice, but two heavy infantrymen attacking one seaman still only add 1, even though this is the same advantage that one heavy would get when fighting one seaman. Two archers attacking, say, a Captain, will add 1 for being two versus one, but so will the Captain add 1 for being of superior rank and equipment, and so the fight is equal. An Admiral, who is assisted, fighting another Admiral will add 1 to his dice, even if his assistance is only from a lowly sailor. If the single man wins the fight he only slays one of his opponents; if they are of the same equipment and rank then the loser may remove whichever one he wishes, but if they are different, or if the winner perceives an advantage in a particular man surviving then he may call for the survivor to be diced for. A 4, 5 or 6 will indicate that the right-hand man of the pair is dead, whilst 1, 2 or 3 is the left-hand man killed.

Two rounds of hand to hand fighting will take place each move until the ship is taken or until the boarders retreat back to their own ship. Between rounds men may be moved about the ship to attack new men or reinforce mêlées.

Repelling boarders and surrendering the ship—For the first move, (two rounds) of hand to hand fighting we assume that both sides are fighting tooth and nail without thought of defeat. After two rounds we start to examine the circumstances under which the combatants may decide either to give up the ship or abandon the boarding enterprise. We can simplify this business of morale to two factors, the strength of one force vis à vis the enemy and the inspiration and leadership given by officers. Count every man aboard the fought-over ship as one point, count the Captain as two points and count an Admiral as three points.

After three rounds of mêlée the boarders will retreat to their own ship if at any time they are outnumbered by the defenders, by three to two if the Captain is aboard with them or, if the Captain is not with them, or dead, they will retreat if outnumbered by any amount. Similarly, the defenders will surrender the ship if after three rounds they are outnumbered two to one. To give an example of how the morale system works we can say that after three rounds, five Roman marines and their Captain are still battling away aboard the Greek Flagship, which is defended by two marines, two archers, two sailors, the Captain and the Admiral. The Roman heavy infantry force is arguably more likely to win a fight to the death, but if we count the points value we will see that 'the lively satisfaction and gratification' which the Greeks feel in being called upon to fight alongside their Captain and Admiral together is paramount. The Romans have five soldiers worth one point each and a Captain at two points; the Greeks have six men at one point, a Captain at two points and the Admiral at three. Seven points against 11, the Romans must retreat.

Archery—Bowmen may either move in a game move or shoot their bows, but not both. Shooting is normally carried out at the end of the move but there is no objection to it occurring at the beginning providing that a full game move has elapsed since the last firing. An archer may therefore, fire at the end of move one and at the end of move two, or at the beginning of move one and at the beginning of move two, or at the beginning of move one and the end of move two but not at the end of move one and the beginning of move two.

The maximum range of a bow is 18 in. Up to 6 in is short range and above that long range. Throw one dice per man shooting, 6 is a hit at long range and 5, 6 at short range.

At long range it is assumed that the archer has simply aimed at a galley and fired and that the arrow may hit anyone aboard; throw a dice for every long-range hit, 1, 2 means a hit on the fo'c's'le (pardon the anachronism) 3, 4 amidships and 5, 6 on the quarter deck (again, pardon). Having decided that a hit is on, say, the quarter deck, number each of the men in that part of the ship and throw another dice to determine which one has been hit. If, for instance, there is a helmsman, a heavy infantryman and the Captain on the quarter deck we might call them one, two and three respectively and throw, say, a 5 on the dice. Counting round we would find that the heavy infantryman was the man hit. There is, of course, no need to count round if all the men in the group are of the same type. When hits are made amidships two additional numbers are included to represent the port and starboard oarsmen.

At short range archers, provided they are firing at the more or less broadside of a ship, may select the area of the target hit, ie, forward, amidships or aft. If the target is presented end-on archers at deck height must fire at the nearest part but archers in raised towers may still choose the part of the ship they fire at.

Archers in raised towers may, at short range, pick out the particular man they wish to kill, eg, the Admiral, or the helmsman, but a 6 is needed on the dice to hit.

That a man has been hit does not necessarily mean that he is killed, or even rendered *hors de combat*. Bows were weak in those days and armour and shields common. For each man hit throw a dice: unarmoured men (ie, archers and seamen), 1, 2, 3, 4—dead. 5, 6—out of action for two moves. Armoured or shielded men, 1, 2, 3—dead. 4, 5, 6—out of action for two moves. Oar sections, 1, 2—men dead, remove one section. 3, 4—half men are dead, remove one section if hit a second time. 5, 6—OK.

Wounding and putting a man out of action for a couple of moves can be quite useful when that man is a helmsman or someone trying to plug a ram hole.

Bolt-throwing engines—These engines, or *ballistae,* were like giant crossbows. We treat them like archers but they only fire once every other move if operated by the normal two-man crew, or every three moves if operated by one man. The range is 24 in, divided into short range, 0–8 in, medium range, 8–16 in, and long range, requiring a 4, 5, 6; 5, 6 and 6 respectively on a dice to score a hit.

A *ballista* bolt was a more formidable thing than an ordinary arrow and so we say that men are automatically killed if hit by one, armoured or not. Oar sections are treated the same as if hit by archers, though.

Stone-throwing engines—The range is 18 in. Up to 9 in is short range, requiring a 5, 6 to hit, and up to 18 in long range requiring a 6 to hit. Throw one dice for each hit: score 6—galley is holed and has taken aboard one load of water (treat as if rammed and attempt to plug the hole); score 5—oar section destroyed; score 4—one man killed forward; score 3—one man killed aft; score 1, 2—no damage.

Stone throwers may fire once in three moves with a two-man crew, or once in four moves with one man.

Morale

When the Captain of a galley is killed it is likely to affect the morale of the crew; he was, after all, not only the virtual owner of the ship but also the employer and paymaster of the crew. When he is killed throw a dice: 1—galley withdraws from action; 2, 3, 4—galley carries on; 5, 6—new Captain is elected.

When a new Captain is elected, we mean that the second in command is sufficiently authoritative to step into the dead man's shoes. Nominate one of the remaining men as the new Captain. By 'galley carries on' we mean that the galley will withdraw if at any time the complement of sailors and soldiers is reduced to half, but otherwise will act as ordered. When the galley withdraws from action it will return in the direction from which it has come, or out to open sea, by the shortest or easiest route. It may not ram or board an enemy whilst retiring, but it may shoot arrows and defend itself if attacked.

If the Admiral is killed it is likely that the fleet would not get to know of this until after the battle, but if his Flagship is sunk or captured the effect of this on the rest of the ships could be disastrous. Throw a dice for each galley next to the sunken or captured Flagship: 5, 6—carry on fighting; 3, 4—ships with Captains (original or elected) carry on fighting; 1, 2—withdraw.

Repeat the dice throw for the ships adjacent to those next to the Flagship but add 1 to the dice if the previous ship's dice throw allowed it to stay.

Chapter 5

The Battle of Thestos c 200 BC

The sea fight described here took place in what was primarily a land campaign in which the two armies of 15 mm model soldiers were predominant, and the fleets merely ancillaries. Despite their secondary role the galleys did in this battle decisively affect the course of the war and so it is worth describing briefly the campaign which led up to it.

The city of Paulopolis, under its Chief Magistrate Haguearchos, was the acknowledged leader of the Euphoric League, a loose alliance of seven city states which formed the Hellenised territory of Euphoria. Euphoria lay in uneasy proximity to the Roman province of Transit Gloriamundi with which it shared a short length of border and which it faced across the smooth waters of the Billeous Sea.

Haguearchos deeply mistrusted the Governor of Transit Gloriamundi, the crafty and cruel P. Hanlius Knutsfordicus—and with good reason. His term of office was due to end shortly and it was his dearly-held wish that his return to Rome should be marked with a Triumph. As his activities so far had barely justified his salary, Knutsfordicus was working overtime to find some area suitable for a conquest to merit him his ovation. His thoughts eventually settled on Euphoria; a fertile land of prosperous cities but politically divided for easy victory. Knutsfordicus' army consisted of 250 men and six ships whilst the forces of the Euphoric League totalled nearly 350 men plus six elephants and nine ships. This mighty assemblage, though, consisted of the separate armies and navies of the seven cities and it was notoriously slow to combine. If he struck swiftly he could conquer the land whilst the declarations of war were still being debated in the city assemblies.

The trouble with the Euphorians was their excessive piety. No major political decision, and especially no decision relating to war, could be made without divine sanction. Each Sunday before the city altars priests would address their city's god with the proposition that the city go to war; they would then cast a sacred numbered ivory cube and interpret a 6 as an Olympian 'aye' and a 5 or less as a divine 'no', in which latter case they would inquire again the following Sunday. The inevitable result of the cities worshipping different deities was that they all went to war at different times and so courted defeat in detail.

The only exception to the pious procedure was the city of Paulopolis where Haguearchos had long ago infiltrated the priesthood of Bacchus, the city's patron god, with friends and relatives and so managed to introduce such liturgical changes as the use of dice with six 6s in religious ceremonies. The city

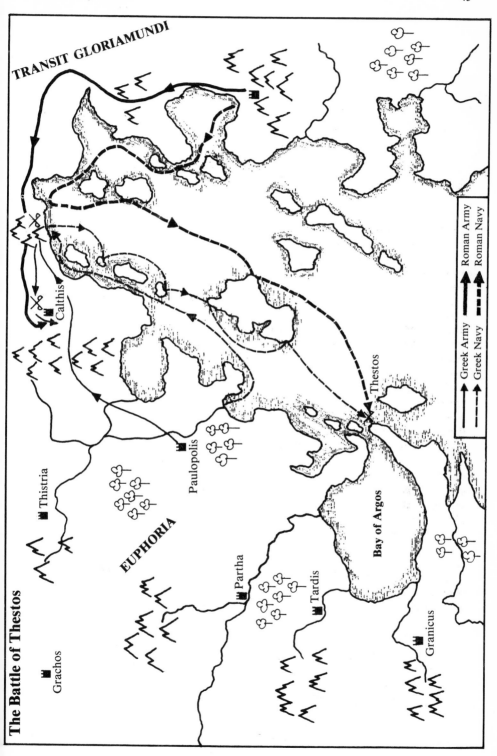

The Battle of Thestos

TRANSIT GLORIAMUNDI

Calthis

Thestria

Grachos

EUPHORIA

Paulopolis

Partha

Tardis

Granicus

Bay of Argos

Thestos

Greek Army
Greek Navy
Roman Army
Roman Navy

could thus mobilise as soon as the political situation required it, but her army of 125 men and six ships, although the biggest single element in the League's forces, would be hard pressed to defend the whole territory until her allies came to their support. If the Romans occupied any city before it had received divine mobilisation orders they would undoubtedly remove the sacred dice from the temple precincts and so prevent that city joining the war.

The war commenced with the Paulopolians holding the mountain pass which joined the two countries, with their fleet protecting their seaward flank against an enemy landing in their rear. The Romans arrived at the pass after seven days and despite a couple of bloody repulses they forced the Greeks to retreat. Nothing so imaginative as the anticipated seaward landing was attempted because the Roman galleys were occupied guarding the supply convoys against possible Greek interference. The battle of the pass had caused gratifying casualties to the Roman Army, but the week's time bought had only resulted in one city, Grachos in the west, declaring war and her army only numbered 30 men. Haguearchos made a stand outside the, as yet, undeclared city of Calthis but he was again defeated and forced to retire. Whilst Knutsfordicus put Calthia to the sack the Paulopolian Army fell back to the mountain pass connecting Calthia with the west of Euphoria. With the war moving inland the Euphorian Fleet was sent south to base itself upon the Island of Thestos which lay at the mouth of the bay of Argos. The closing of this bay to enemy ships was of critical importance as feeding into it were three navigable rivers giving access to three powerful and, as yet, undeclared cities whose fall would lead inevitably to a Euphorian defeat.

The Roman Fleet continued for a few days longer to guard its armies' supply lines until this became unnecessary. Only then did Knutsfordicus' thoughts turn to an attack by sea on south Euphoria. The fleet was eventually despatched to the bay of Argos with its primary objective being the capture of the city of Tardis whose army of 65 men and four elephants, and navy of three galleys made her the second strongest city in the League. The Romans took five days to reach the bay entrance south of the Isle of Thestos, sailing when the wind was with them and rowing when it was not. As they approached to within sight of the island a beacon flared on a cliff top. It was a Paulopolian lookout station and the flames were a warning to the Greek commander to launch his ships from the beach on which they lay.

The ubiquitous P. Hanlius Knutsfordicus was commanding his fleet in person and he had to make up his mind whether to fight there and then with his oarsmen tired and those of the enemy obviously fresh from resting, or whether to retire to some quiet cove and there let his rowers recover from the efforts of the journey. He decided on the first course of action; already a second city had thrown in its lot with Paulopolis and soon there would be more. If he was to attack the cities of the south he would have to do it quickly whilst they still prevaricated. His fleet sailed on, into the narrow Straits of Thestos.

*   *   *

The Roman Fleet consisted of six war galleys, all equipped with fighting towers and two of them were also equipped with *corvii*. They were indifferent seamen (and turned on circles one size larger than usual) but relied on large boarding crews for victory, each ship having two archers and six armoured marines, in addition to the two sailors and Captain. The ships were: *Amo* (equipped with *corvus*), *Amas, Amat, Amamus* (Flagship, equipped with *corvus*), *Amatis, Amant.*

The Greek commander Haguearchos of Paulopolis, had also managed to whisk himself from the mountain pass in the north to lead his fleet on the day. The Euphorian Fleet was crewed by prime seamen who relied on ramming skill to win their battles, consequently only four heavy infantry and two archers were carried aboard its galleys, as well as the Captain and two sailors. Haguearchos' Flagship, the *Gamma,* mounted a bolt-throwing *ballista* in its bows, but none of the ships had either *corvii* fitted or fighting towers. The Greek ships were named thus: *Alpha, Beta, Gamma* (Flagship), *Delta, Epsilon, Zeta.*

The Straits of Thestos were very narrow at one point but widened out rapidly on both the bay and seaward sides of the island tip. Haguearchos had decided that with the great rowing and manoeuvring powers of his ships the most advantageous place to fight was in the narrowest part of the strait. Quite how he had arrived at this fallacious notion is still one of the mysteries of history but by basing his plan of battle on this false concept he played straight into the hands of the enemy whose 'close and board' tactics were ideally suited to a congested action.

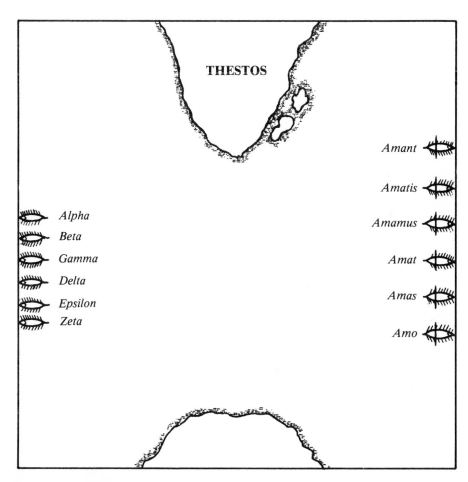

Moves 1–4—The Greek Fleet arrayed itself in a compact line abreast and advanced slowly from the bay end of the straits towards the narrows whilst the Romans sailed towards them from the seaward end, frantically taking their sails in and lowering their masts.

The Greek advance conformed with that of the Romans and when the Romans hove-to for a couple of moves to complete the stowage of masts and sails the Euphoric galleys followed suit. Haguearchos was not going to allow himself to be lured into the perils (?) of wide waters if he could help it.

Move 5—Haguearchos at last had an idea! The enemy ships were more widely spaced than his own and he rather thought that if he managed to edge his fleet past the end of the enemy line so that some galleys were before it, some behind it and some alongside it he would be able to achieve a tactical concentration on that flank.

All this required a supine inactivity on the enemy's part, of course, but what else could he do? The first suggestions of doubt regarding his chosen battle-ground were just beginning to gnaw at his buoyant self-confidence.

In accordance with this vague substitute for a plan the galleys of the Greek Fleet all moved forward and turned 45 degrees to starboard preparatory to forming a line ahead. Preparatory, because only then was it realised by he who should have realised before that the ships were too closely packed to allow the change of formation without a deal of heaving-to and backing water.

The Roman line rowed on at slow speed.

Move 6—The leading Greek galleys increased the pace to cruising speed and turned to head the line along the mainland shore of the channel. The rearward galleys somehow managed to jostle themselves into a ragged single file.

The three right-hand ships of the Roman line, *Amamus, Amatis* and *Amant* increased to cruising speed whilst the left-hand three continued rowing slowly.

Move 7—While the Greeks pulled on at cruising speed in their untidy L-shaped line ahead, the right-hand Roman squadron increased to fast speed and tore down upon the Greek rear, whilst their left-hand squadron put their helms over and turned rapidly towards the Greek van.

The range had closed considerably and the archers in both fleets loosed their arrows at short and middling ranges. The *Amo* had one man killed by a *ballista* bolt from the Greek Flagship, and two men wounded and put out of action for two moves, by arrows from *Zeta* and *Epsilon,* whilst one of *Epsilon*'s archers was killed by return fire from *Amas.* In the rear an arrow from *Alpha* wounded a man aboard *Amamus.*

Move 8—The Roman galleys of the left squadron sped in to ram the Greek van. *Epsilon,* second in the Greek line, turned sharply to starboard to avoid the *Amo*'s beak but the Roman was almost as quick with his helm and struck her at a shallow angle on the starboard side. By so doing she brought herself across the path of the *Delta,* third in the line, and so was inevitably rammed amidships. *Amo*'s starboard, though, was covered by *Amas* and so as *Delta* shuddered to a halt against *Amo* she was ripped with *Amas*' ram. Meanwhile *Zeta,* the leading Greek, had turned tightly to port and was now bearing down on the *Amas* herself at cruising speed.

The Roman archery in this part of the battle was quite ineffective but one archer from *Epsilon* hit a wounded marine on *Amo,* killing him, and *Delta*'s archers killed an archer in *Amas*' tower.

1 *Amo* rams *Epsilon* destroying one oar section. Cruising speed, angle B. Dice

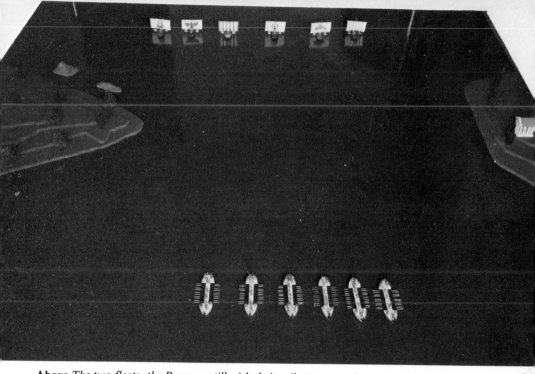

Above *The two fleets, the Romans still with their sails set, enter the strait.*

Below *The Greek Fleet angles to the right. In the centre can be seen the Flagship* Gamma *with a* ballista *in the bow and Haguearchos in the stern brandishing his sword heroically.*

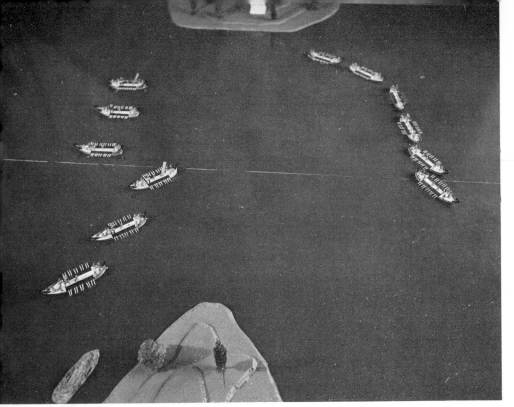

Above *The galleys of the Roman right flank spurt forward as the Greeks jostle into line ahead.*

Below *The Romans bear down on the Greek van and rear and archery commences.*

Above *The Roman left and the Greek van engage in a ramming* mêlée.

Below *The* Amamus' *marines capture the* Alpha *whilst an attack by* Beta *is defeated. The cloaked figure of P. Hanlius Knutsfordicus can be seen on the Romans' quarter deck. On the left of the photo* Amant *drifts with her entire starboard bank of oars destroyed.*

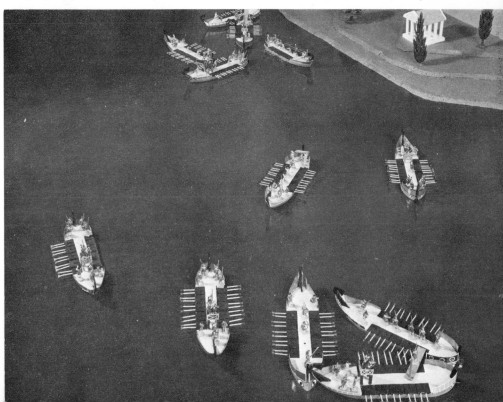

throw 3 + 3 = 6, – 1 (angle B) = 5, no hole. **2** *Delta* rams *Amo,* destroying one oar section. Fast speed, angle B. Dice throw 5 + 2 = 7 + 1 (fast), – 1 (angle B) = 7, holed and one load of water taken in. **3** *Amas* rams *Delta* at cruising speed, angle A, destroying an oar section. Dice throw 4 + 3 = 7 = holed and one load of water taken aboard.

At the rear of the line as well as in the van the Romans held the initiative. *Amamus* sped into a ramming attack against *Alpha;* the Greek turned to starboard to avoid but as *Amamus* cut inside they collided cheek to cheek by the bows and the Roman dropped her *corvus* locking the two ships together. The armoured marines swarmed across the boarding bridge and after two rounds of mêlée they had lodged four men on the enemy's deck killing two *hoplites* and an archer with no losses to themselves.

Beta turned sharply round to port hoping to place herself in line to ram but without success as the Romans were moving and turning too fast, two of her men, though, were wounded and knocked out for two moves by *Amatis'* archers. *Gamma,* with Haguearchos aboard, was luckier; turning around by 180° to port she made a ramming attempt against *Amant* who had turned back to reinforce the van. The ramming attack failed as she took the enemy at too shallow an angle but she raked her along the starboard side, shearing all her oars on that side.

Move 9—Aboard *Delta* up in the van of the battle the seaman allocated to the job managed to block the hole in her side (ie, 5 on the dice) and as she backed water to withdraw her ram from *Amo* one of her archers shot the helmsman aboard the Roman. The *Amo*'s damage repair party was not so successful and a second load of water flooded into her hull. Encouraged by the number of casualties to bow fire aboard the Roman, the crew of the *Epsilon* threw grappling lines across her rails (two held) and boarded. They slew the marine lying wounded in the bows but the ferocity of the Roman defence was such that two *hoplites* and the Greek Captain were killed and they withdrew back to the *Epsilon.*

As the *Amas* backed water to extricate herself from *Delta* the *Zeta* rammed her full and square at fast speed in the port side. A dice throw of 5 + 4 = 9 + 1 (fast speed) = 10 = holed, four loads of water taken aboard and an oar section destroyed.

In the rear of the battle *Beta* moved to *Alpha*'s support by running alongside *Amamus'* stern, grappling her and boarding. Only one marine, the Captain and two sailors were at the Roman stern but they fought with such savagery that the three *hoplites* and one archer who gained a foothold on the quarter deck were all killed without loss to the defenders. Up in the bows the Roman marines, unperturbed by the activity behind them, continued in their desperate fight for the *Alpha;* again they outfought the Greeks and killed *Alpha*'s Captain, two *hoplites,* and a seaman. The remaining seaman and archer surrendered and gave up the ship.

The crippled *Amant* lay drifting as her oars were redistributed on both sides of her hull. The Greek Flagship *Gamma* turned about to starboard with the idea of ramming her but on seeing *Amatis* approaching she thought better of it. One of her *ballista* crew fell dead with an arrow from *Amant*'s tower.

Move 10—*Zeta* pulled back from the *Amas* who, failing to block the ram hole in time, took a fifth load of water aboard and capsized. Her seamen and soldiers were diced for and five of the 11 managed to divest themselves of equipment and swim for safety.

The *Amo* was more fortunate as on this move her men were able to stop the leak, although with two loads of water in her hull she would be reduced to half speed. Up on deck *Amo*'s marines followed up the repulse of the Greek attack by dropping her *corvus* on *Epsilon*'s deck and counter-boarding. With the amazing luck in hand to hand combat which the Romans had enjoyed throughout this battle they killed two *hoplites* and two archers without loss to themselves, and faced with only two sailors remaining they took the ship.

The *Beta* having had enough of cold steel for one day cut her grappling lines, pushed off, and rowed away from *Amamus*. The *Amat,* which had been rowing up to support *Amamus* in her boarding action ran alongside *Alpha*. As she was surrendered it was too late to give assistance but she was able to provide a seaman and soldiers to stand as prize crew aboard her.

The *Amatis* increased the pace to fast speed and went after the Greek Flagship, which was only moving at cruising speed and so was quickly overhauled. Deft use of the *Gamma*'s helm prevented her being rammed but *Amatis'* bow struck her stern rails, carrying away her starboard rudder and allowing her to be grappled. The archers up in *Amatis'* tower fired down on the Greek quarter deck, one aiming at Haguearchos himself but missing, and the other at the Captain, succeeding in wounding him. The marines attempted to board but for once the Greeks fought back effectively and two soldiers on each side fell.

Move 11—The *Delta* slowly got under way again and turning her bows nosed

In the middle ground Amatis *and the Greek Flagship are locked together in a boarding fight. Beyond them is the sinking* Amas.

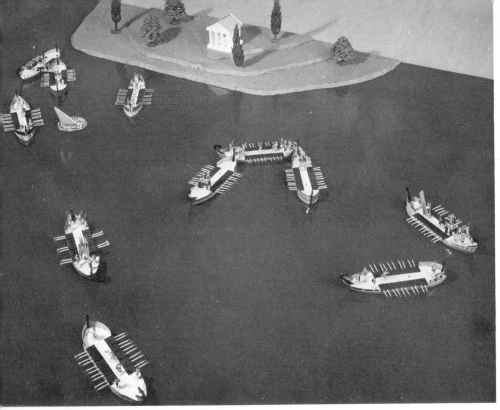

Above *The Greek Flagship is taken.*
Below *The Greek survivors flee the field.*

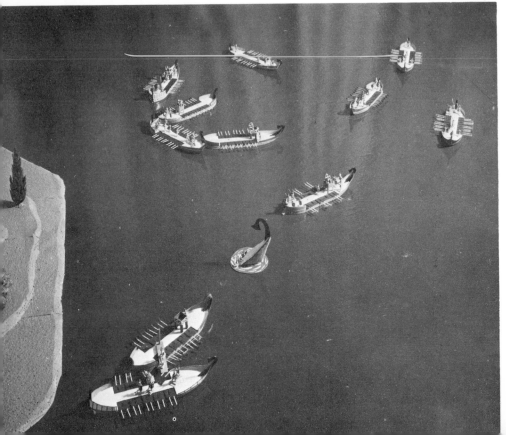

her way towards the Flagship.

The Roman archers aboard the captured *Epsilon* fired at *Zeta* which lay still after backing water the previous move, and killed an archer. *Zeta*'s archers replied but unfortunately hit one of the Greek prisoners aboard *Epsilon*.

Amat, backing water away from the captured *Alpha,* went stern-first against *Gamma*'s bow. Making the most of the opportunity the Romans threw grappling lines, but despite only one catching fast they boarded the Flagship and ran the risk of their own ship drifting away and leaving them without support. The marines being in the *Gamma*'s stern fighting the *Amatis'* crew, the only opposition to the *Amat*'s assault were the archers, seamen and *ballista* operator; they killed three of them for one soldier of theirs. In the stern the two remaining *hoplites* and Haguearchos were killed after slaying two Roman marines. With the Admiral dead, the Captain lying wounded and the heavy infantry all dead, the two remaining crew were completely outnumbered. They surrendered.

Move 12—The two Greek galleys seeing the surrender of the Flagship, *Beta* and *Delta,* tested their morale. Both threw a 2 and so had to break off action. *Zeta* moved forward and also saw the surrender; she threw a 5 and so was at liberty to carry on fighting, but not relishing taking on the entire enemy fleet single-handed her Captain adopted the discreet course of action rather than the valiant one and fled with the other two.

Having much work left in their oarsmen the Greeks were able to leave the scene of action at high speed; the Roman crews were all near to exhaustion and so unable to pursue.

The battle was a decisive Roman victory. With the loss of only one galley and 17 men killed or drowned, they had captured three Greek ships, killed 28 men and taken eight prisoners. Much of Knutsfordicus' success was attributable to an unbroken run of lucky dice throws in the boarding fights, but it cannot be gainsaid that he fought the battle with an aggression and a determination that contrasted markedly with the half-heartedness and tactical ineptitude of the Greeks.

Postscript—With the battle won the Roman oarsmen were rested for a day whilst the seamen carried out necessary repairs to the ships. The fleet then pressed on with its prizes in tow and on reaching the city of Tardis overawed its citizens and secured its non-participation in the war. A similar operation against Granicus was narrowly defeated by a declaration of war when the fleet was less than a day's journey from the city, but the fall of Tardis was sufficient to ensure an ultimate Roman victory and a Triumph for Knutsfordicus on his entry into Rome. Sad to relate the men of the fleet were unable to take part in the celebration as, when the war was drawing to a close, the retreating Euphorian Army came upon them with their ships beached and an inadequate watch set, and they were slaughtered almost to a man.

Chapter 6

Napoleonic naval warfare

Perhaps this chapter would have been more appropriately entitled 'Nelsonic naval warfare' as it is around the sea battles of Nelson and his contemporaries, the American War of Indpendence and the French Revolutionary and Napoleonic Wars, that the rules in this chapter are cast. This pitching of them at the end of the 18th century does not make them very exclusive to this period, though, because in the 250 years which separated the Spanish Armada from the installing of steam engines in ships of the line, sea fighting changed so little that these rules can be used, with hardly any alteration, for naval wargaming in any of the periods of gunpowder and sail.

Tactics

The 'wooden walls' mounted almost all their artillery so that it fired on a narrow arc on the broadside and consequently the old fighting formation of line abreast disappeared in Elizabethan times in favour of the line ahead. The line ahead was initially an ad hoc formation occurring in groups and squadrons, brought about by the need to keep the guns' field of fire clear of friends, but by the beginning of the 18th century it became *de rigeur* for the entire fleet to form one line.

The original purpose of the notorious *Permanent Fighting Instructions'* insistence upon a single line ahead formation for all circumstances (save a chase) was to ensure that the battle did not get out of the Admiral's control, as had often occurred during the more pell-mell mêlées of the Dutch Wars, but by also insisting that the line be countermineous to the enemy's they made a decisive victory almost impossible to achieve. 'Countermineous' means that the line ahead must be drawn up with the van facing the enemy's van, the centre his centre, and the rear his rear, thus defying the cardinal rule of tactics—concentration at the decisive point. It is because of the *Permanent Fighting Instructions* that whilst there were a great many sea battles during the first three quarters of the 18th century there were very few decisive ones.

During the 17th century sea wars against the Dutch, before the *Permanent Fighting Instructions* had appeared, there were developed a number of methods of concentrating one fleet against another and these were revived in the latter quarter of the 18th century when Admirals were given greater freedom in the conduct of battles. These tactics may seem absurdly simple, but the difficulties of communicating from a shot-torn deck meant that they were difficult to achieve in practice and they will be equally difficult to achieve in the wargame if

Massing

Admiral Black has massed against Admiral White's rear and so although outnumbered he can bring the fire of seven ships against five of the enemy.

the rules are properly applied. This means, in effect, that you must out-think your opponent before the battle even commences and that was the very art of the sailing Admiral.

The first method of concentration is massing—a tactic employed with success by Rodney at the Battle of Martinique. The fleet sails parallel with, and on the same course as, the enemy in line ahead but with a smaller spacing between ships. As a result a greater number of the ships in the massed fleet can fire against a fewer number in the opposite line. Wargaming experience of massing suggests the following difficulties in its adoption; firstly the limited arcs of traverse of broadside guns means that the masser may be forced to fight at a longer, and consequently less decisive, range in order to bring all his guns to bear. Secondly, he is likely to find that the enemy are reasonably spaced so that his ships when massed will be dangerously close together—and his opponent will be watching like a hawk for collisions.

An alternative method of achieving local superiority against an enemy line is by doubling it, which means that the attacking fleet is formed into two squadrons which position themselves on each side of the enemy line. The enemy ships between the two squadrons are subjected to a murderous crossfire whilst those which are not are unable to engage. Nelson employed doubling at the Battle of the Nile, but this was made easier to achieve by the fact that the French line lay at anchor.

Doubling

Concentration is easier to achieve against a moving enemy by approaching his line at right angles and cutting it with your own. If the wind direction is favourable this will give the effect of isolating his van and allowing you to thrash his centre and rear with your whole fleet whilst his leading ships are still trying to beat back against the wind to join the fight. This is basically the tactic employed by the British Fleet at Trafalgar and you cannot ask for a better recommendation than that. Its principal disadvantage, though, is that the attacker is subject to raking fire during his approach which he is unable to reply to.

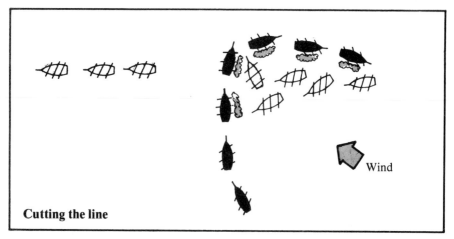

Cutting the line

Admiral Black has cut the White line and achieved a local superiority against the disorganised centre and rear. The White van will have to beat back against the wind and by the time that it joins the fight, Black hopes, the rear and centre will be well beaten.

The gaining of the windward position, or the 'weather gauge' as it was called, was an important part of aggressive naval tactics in the days of sail. The fleet which had the weather gauge could run down on the enemy and force them to action in a way that a leeward fleet could never do, and so often hours or even days would be spent before a battle by the rival fleets manoeuvring for the advantageous position.

It became standard practice in those navies with an offensive policy always to seek the weather gauge so that a decisive battle could be ensured. The French Navy during the late 18th century, however, adopted a defensive role. Their fleets would fight for some specific objective, say the covering of the passage of a convoy, but they would not fight for the sole purpose of destroying the enemy as did the Royal Navy. Consequently their fighting system developed along different lines, into the doctrine of tactical evasion whereby the battle was fought from leeward to facilitate an easy downwind retreat when the danger to whatever it was that the fleet was protecting had passed. To ensure that the withdrawal would not be interfered with their guns were fired, not into the enemy's hulls, as was British practice, but into their masts and rigging so that they would be too crippled to pursue. These opposing policies of upwards and downwards firing were further encouraged by the angles of the gun decks of the rival fleets in their upwind and downwind positions.

The problem which British Admirals were faced with was how to prevent the enemy running away when your masts and yards were too shot about to allow you to chase them. Lord Howe overcame the difficulty at the Glorious First of June (1794) by attacking from windward and so forcing the French to action, and then breaking his line at all points and fighting them from leeward, thus blocking their retreat. The manoeuvre by its very nature brought about a mêlée of single-ship actions in which concentration was lost. This did not matter to Howe, whose ships were individually more efficient and faster-firing than the French, but in a wargame between equal opponents the result might be a slogging match. In one or two wargames of my own, however, where this

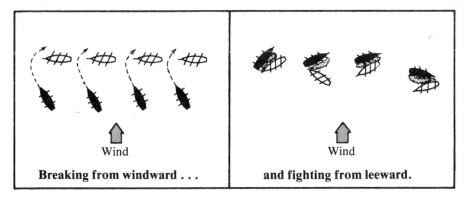

| Breaking from windward . . . | and fighting from leeward. |

stratagem has been used and the leeward fleet has been firing only into the masts, the windward attackers have so reduced the enemy's fire-power by firing into their hulls that after breaking the line they *were* individually superior so that the attacks were successful.

The rules

I like my model men-of-war to display some of the damage they have suffered with removable masts and so I use the Airfix *Victory* kits, cut down and converted, rather than the otherwise more convenient 1:1,200-scale models of ships of this period which I have always found to be a bit too small and fiddly to have anything other than fixed masts. This, then, accounts for the distance scales and almost all-or-nothing system of mast damage which I have adopted in these rules. If you prefer to use the smaller-scale models the distances can be halved to allow the battles to be fought on a large table top and speed reduction can be made more progressive.

The difficulty of using the larger-scale models is the amount of floor area they occupy, but pre-battle manoeuvres, such as scrambles for the weather gauge, are best plotted out beforehand on a map, whatever scale of model is being used.

Sequence of play

1 Decide upon movements to be made and the types of shot being used: note them down on an order sheet and when both players are ready, declare them.

2 Move models up to the maximum distances allowed, both players moving simultaneously. Where it appears that ships may collide phase moves are used.

3 Guns are fired. All firing is deemed to be simultaneous so that a ship which is damaged in a move may still return fire at the full rate.

4 Boarding and boarding mêlées are fought out.

5 Dice are thrown for any ships in danger of striking.

Movement

The speed of a sailing ship is no absolute quantity but one which varies with a host of factors, the most important of which are wind strength and the point of sailing (ie, the course of the ship in relation to the wind direction).

The wind

If the battle is part of a campaign or one in which the earlier manoeuvres have been plotted on a map the wind strength and direction will be already known. If the battle is a one-off the tables below can be used to determine them. Wind direction should always be diced for as it confers tactical advantages but it is usually best simply to agree upon a moderate or fresh wind strength as it can be annoying for two players to come together for a wargame only to find themselves becalmed, or scattered by storm.

For wind direction throw a dice as described on page 35. Throw a dice every third move to see whether the wind direction changes. 6, wind veers by 45 degrees (ie, clockwise); 5, 4, 3, 2 wind direction constant; 1, wind backs by 45 degrees (ie, anti-clockwise).

An Admiral may base his plan of attack on the assumption of a prevailing wind direction, say on-shore and off-shore breezes at different times of the day. In such cases we want the choice of wind direction to be still variable but biased in the prevailing direction. When, therefore, the wind is blowing in the 'right' direction use the wind change table given above, but when it is blowing in a different direction throw a dice every three moves but alter the direction as follows: 6, 5, 4 wind backs or veers 45 degrees towards the prevailing direction; 3, 2, wind stays constant; 1, wind backs or veers 45 degrees away from the prevailing direction.

To determine the wind strength throw a dice and read off from the column of the appropriate season.

	Winter	**Spring/Autumn**	**Summer**
6	Moderate	Light	Calm
5	Fresh	Moderate	Light
4	Fresh	Moderate	Moderate
3	Gale	Fresh	Moderate
2	Gale	Fresh	Fresh
1	Storm	Gale	Fresh

Sailing

In order to combine wind strength and direction into a speed for a ship we give each ship type a move which varies with the wind speed, and then multiply it by a factor for its point of sailing.

Movement factor table

			Wind		
Ship type	**Light**	**Moderate**	**Fresh**	**Gale**	**Storm**
Ship of the line	1 in	$1\frac{1}{2}$ in	2 in	$1\frac{1}{4}$ in	$\frac{1}{2}$ in
Frigate	$1\frac{1}{4}$ in	$1\frac{3}{4}$ in	$2\frac{1}{2}$ in	1 in	$\frac{1}{2}$ in
Sloops, cutters &c	$1\frac{1}{2}$ in	$1\frac{3}{4}$ in	$2\frac{1}{4}$ in	$\frac{3}{4}$ in	Nil

Once the wind strength has been decided upon the movement factors remain constant throughout the battle.

Point of sailing factors are as follows:

Running, ie, wind behind the ship— 4

Broad reach, ie, wind on the rear quarter— 5

Reach, ie, wind at 90 degrees to the ship's course— 3

Haul, ie, the ship sailing into the wind (not closer than 45 degrees)— $1\frac{1}{2}$

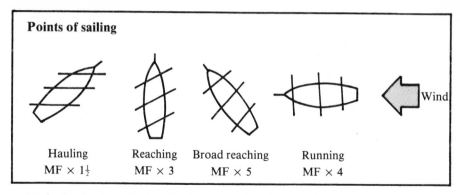

Points of sailing

Hauling	Reaching	Broad reaching	Running	Wind
MF × 1½	MF × 3	MF × 5	MF × 4	

At the beginning of a battle, then, we simply note the movement factors for the prevailing wind strength for each type of ship, multiply them by the four point of sailing factors and those are the speeds, subject to damage, at which the ships will sail for the rest of the battle. For example, a frigate in a moderate wind has a movement factor of $1\frac{3}{4}$ in and so will sail at 7 in running, $8\frac{3}{4}$ in on a broad reach, $5\frac{1}{4}$ in on a reach, and $2\frac{5}{8}$ in hauling.

Calms

In a 'calm' individual ships may still be able to move, by catching odd gusts of wind. Throw one dice per ship, 5, 6 indicates that a gust has been caught and that the ship may move. If the ship moved last turn on a gust then 4, 5, 6 indicates that it may still move. On these occasions the ships will move with a factor half that of the light breeze factor.

Turning

Three turning circles are required: ships of the line—4 in diameter; frigates—3 in diameter; brigs, sloops, etc—2 in diameter.

Normally, when a ship is changing direction it will simply be moved around the turning circle and then continue on its new course. The only variation to this is when the turn would cause the ship's bow to cross the 'eye' of the wind. Under these circumstances the ship must either 'wear' or 'go about'.

Going about

A ship may only go about in light or moderate winds and when it has got all its masts standing. Ships which have lost a mast, or in fresh or stronger winds, must wear to change tack.

Going about involves the ship swinging its bow 90 degrees around its stern to bring it on to the opposite tack. This process takes a whole move during which period the ship does not move forward. A ship must haul for at least another two moves before it can go about again.

Wearing

When a ship is 'worn' it changes tack by presenting its stern to the wind and turning through 270 degrees. The appropriate-sized turning circle must be used in this manoeuvre.

Effect of mast damage

In practice a ship would lose speed gradually as its sails were holed, shrouds

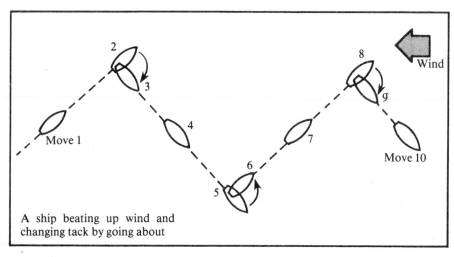

A ship beating up wind and changing tack by going about

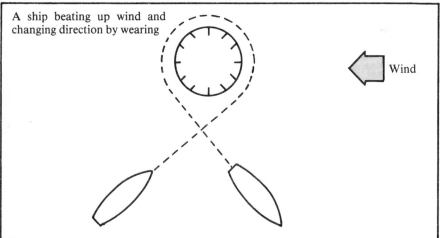

A ship beating up wind and changing direction by wearing

parted and spars and upper masts brought down. As, however, my models have three removable masts the rules are made to conform to this limitation.

When a mast is brought down, the following will happen:

1 The mast will fall to leeward and halt the ship for one move until the wreckage is cut away.
2 The ship's speed will thereafter be reduced proportionately, so that a three-master will move at two thirds speed when it has lost one mast, at one third speed after losing two masts, and will drift downwind at half its movement factor when it has lost all three masts.
3 A ship which has lost a mast may not sail closer than 60 degrees to the wind.
4 The lee battery will be partially masked and will fire at half-effect for that move until the wreckage is cut away.

Falling aboard

A ship whose hull comes into contact with another is said to have 'fallen aboard' her. This may happen accidentally, as a result of clumsy manoeuvring, or deliberately, if a boarding attack is being made. As the rigging of the two ships would become entangled and because a boarding attack would include the use of grapnel lines, separation is not just a matter of sailing off in separate directions.

1 If both Captains wish to separate, a dice is thrown at the beginning of each move. A 5 or 6 indicates that the ships are separated, 1, 2, 3 or 4 that they remain locked.

2 If either of the Captains wishes to board the ships will remain locked together until the boarding mêlée is decided.

Ships which are locked together will drift downwind at the basic movement factor of the slowest of the pair.

Damage record cards

The damage record cards to go with these rules are very simple as all we are going to do is accumulate damage points caused by gunfire and reduce firepower and crewmen available for boarding by a proportionate amount.

We give each ship a hull value of five points for every gun carried, so that a 64-gunner is valued at 320 points and a 100-gunner at 500 points. I value merchant ships as 12-gunners and big East Indiamen as 32-gunners.

It would be tedious to have to re-assess fire and boarding power every time that a point's worth of damage is caused so we tot up the damage suffered and then reduce firepower and the boarding crew by one fifth for every fifth of the hull value reduced by damage. The 'wooden walls', though, were very rarely sunk by gunfire alone so that when a ship has suffered damage points equal to its hull value, and so lost all its firepower and boarding crew, it is still considered to be afloat and it must suffer the same number of damage points again before it finally slips below the waves.

The chart we use consists of six boxes in which hull damage points are accumulated. The first five will have a fifth, two fifths, and so on, of the hull value written in the lower right-hand corners and the sixth will have double the hull value written in it. By the side of the first five boxes are the firepower and boarding factors for the ship in its varying stages of damage. Any damage points inflicted on the ship will be written in the boxes and as the total number of points exceeds the hull value written in the corner the box and fire and boarding factors are crossed through and the next line proceeded to.

Three additional boxes are given for the accumulation of mast damage. As the value of each box is reached a mast falls and speed is reduced accordingly. The value given to each mast is 80 points for ships of the line, 60 points each for those of a frigate and 40 points each for sloops, and other lesser vessels.

Calculation of firepower factor

Multiply the total number of guns of one side, of each size by the weight of shot and divide the total by 20. As we ignore chase guns in these rules the number of guns of any size on a broadside can be taken to be half of the total.

To take HMS *Belerophon* as our example, she has a total of 28 32-pounders, 30 24-pounders and 16 9-pounders. We halve the totals to get the number for one broadside and the calculation is as follows:

$$\frac{(14 \times 32) + (15 \times 24) + (8 \times 9)}{20} = \frac{880}{20} = 44 \text{ firepower points.}$$

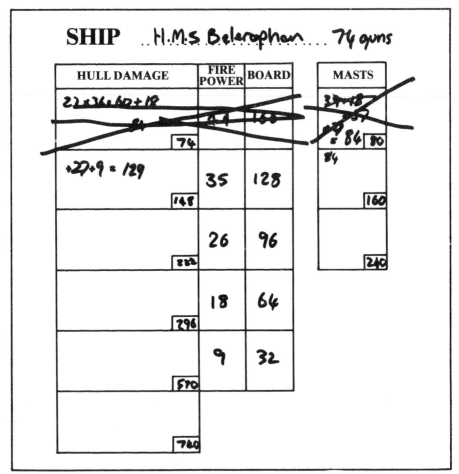

A damage record card for HMS Belerophon. She has suffered 129 damage points to the hull and 84 to the masts, so her firepower and boarding factors are reduced by a fifth and she is down to two thirds speed.

Calculation of boarding factor

The greater part of a vessel's crew during battle would be busy working the ship or firing the guns and so we assume that only a quarter of the crew are immediately available for boarding or repelling boarders. The boarding factor for a ship, then, is its total complement divided by four. By noting down an instruction to that effect on the order sheet two moves in advance, one or both broadsides can be stripped of their crews and these men added to the boarders, in which case the boarding factors will be doubled (one broadside) or trebled (both broadsides). During the two-move assembly period (and the two moves to return to their posts) the stripped broadside may not fire and the men may not fight.

Gunnery

Prior to each move a note must be made on the order sheets of the type of shot that each ship's guns are loaded with. In the absence of any such note the guns are assumed to be loaded with the type of shot used the last time the ship fired, or if the ship has not fired before, single round shot.

The types of shot to choose from are:

Chain shot—used for firing into masts and rigging. If fired into a ship's hull it only causes half-effect damage.

Round shot—used for firing into hulls. If fired at masts and rigging it only causes half-effect damage.

Double round shot—only loaded before the battle commences; once the double shot has been fired single rounds are loaded. Double shotting causes great damage at short range but loses its effect at long range.

Broadside guns may not traverse more than 20 degrees. To ease the business of determining how many guns can and how many guns cannot bear on a target we split the broadside into two, from the foremast to the main mast and from the main mast to the mizzen. If the whole of the broadside cannot be brought to bear on a target but one half of it can then the ship may fire, but at only half effect.

Battleships and frigates carried two or four heavy guns ahead and astern to give fore and aft fire but we ignore these 'chase guns' in this set of rules so that a ship which is raked by another not only suffers greater damage (because the shot travel the whole length of the ship) but is also unable to return the fire.

Firing procedure

Although the sizes and power of guns carried at sea varied enormously, the firing of them from heaving decks and using only crude peep sights was so erratic that they were unlikely to hit anything above short range. Consequently we give all ships, regardless of the sizes of guns carried, the same range of 24 in. Up to 6 in is short range, 12 in is medium range, 18 in is long range and 24 in extreme range. One may not fire at masts at less than 3 in range as the elevation would be insufficient.

Declare the type of shot being fired and whether masts or hull are the target, then measure the range from main mast to main mast. The firer is committed by his declaration so that if he finds on measuring that the target is outside his

Left Black may only fire half his broadside at White as the target ship is only partially within the arc of the forward half of the broadside. **Right** *Black is raking White's stern and so causes additional damage as the shot travels the whole length of the target's decks. White is unable to reply.*

range he must still fire and, if necessary, lose his carefully loaded double shot. If a broadside is loaded with chain shot and the masts are declared as the target, but on measurement the enemy proves to be less than 3 in away, the chain shot will hit the hull.

When the guns are fired two average dice are thrown, to simulate the luck of war, and their scores added together. Points are added to or subtracted from this total as given below:

Range—deduct 1 at medium range; deduct 2 at long range; deduct 3 at extreme range.

Double shot—add 2 at short range; add 1 at medium range.

Raking—add 2 if firing at the hull stern; add 1 if firing at the bows; add 1 if raking the masts.

Gale force winds—deduct 1 from weather side batteries; deduct 2 from lee side batteries.

The total of the two average dice plus and minus any other points is checked on the fire effect table below:

Fire effect table

Dice ± points	Multiply firepower factor by
12	3
11	$2\frac{1}{2}$
10	2
9	$1\frac{1}{2}$
8	$1\frac{1}{4}$
7	1
6	$\frac{3}{4}$
5	$\frac{1}{2}$
4	$\frac{1}{4}$

To illustrate the way in which the firing system works we will take HMS *Belerophon* in the damaged state in which we left her on page 64. Her firepower has been reduced to 35, and she is firing double shot, at medium range, at a Frenchman whose bows she is raking. Two average dice are thrown and they score 3 and 2.

Dice total	5
Double shot, medium range	+1
Raking bows	+1
	7
Firing at medium range	−1
	6

A total of six means that the firepower factor must be multiplied by three quarters, giving a total of 26 damage points inflicted on the Frenchman's hull.

The shot penetrating the hull may also damage the masts by striking them at their bases. Every time the hull is fired at and damage caused throw an additional (ordinary) dice: 6—three quarters of the damage points caused are also inflicted on the masts; 5—half the damage points caused are also inflicted on the masts; 4—a quarter of the damage points caused are also inflicted on the masts; 1, 2, 3—no mast damage.

If, in the example of the *Belerophon* above a 4 had been thrown on the ordinary dice a quarter of 26, or seven points, would have been added to the mast damage score of the target ship.

Boarding

Each ship has its boarding factor which, as I have explained, can be doubled or trebled if gunners are called up with two moves' notice. If both sides are claiming the right to board, the initiative is presumed to be his who has either noted an order to board or who has called up guncrews; if the matter is still undecided let each player throw a dice and the higher score will confer the right.

Each ship is assumed to have one more deck than it has gun decks, so that a first-rater will have four decks, a frigate two, and so on. A ship is taken when the boarders have captured all its decks. Should the boarders be driven back to their own ship the successful defenders may either attempt to separate the ships, or they may counterboard. The boarding mêlée is fought out thus:

1 Each side throws two average dice.
2 The side which boarded this move, or which advanced a deck last move, adds 1 to the total of the dice.
3 The dice totals are multiplied by the boarding factors.
4 If one side has a product (dice × factor) 33 per cent greater than that of its opponent it will advance by one deck. With a product 100 per cent greater, advance by two decks. With a product 150 per cent greater, advance by three decks. With a product 200 per cent greater, accept the enemy's surrender.

If less than a 33 per cent superiority is scored the fight will continue on the same deck. The fight is played out every move until either the boarder has advanced through all decks, and so taken the ship, or until the defenders have pushed the boarders back, deck by deck, until they are back on their own ship.

One or two anomalies stand out in this system of computing the effect of boarding, principally the fact that no casualties are caused by all this cutlass crossing, the sides just push back and forward like a scrum, and no account is taken as the fight advances on to a new deck of the men to be found operating guns there who might reasonably be expected to join in. The method does work quite realistically, though, and it does ensure that boarding fights, which can become impossibly complicated if the rules allow them to be, do not take up too much playing time.

Striking

A vessel may be called upon to 'strike its flag' (surrender) when it has suffered two fifths hull damage (ie, two boxes crossed out). The ship demanding its surrender must satisfy the following demands:

1 Its remaining firepower must be at least 50 per cent greater than that of the surrendering ship.
2 It must have suffered less damage.
3 It must be within 4 in.

A dice is thrown to decide whether or not the ship surrenders. If it carries on fighting it will be re-tested every time a box is crossed out. During the Napoleonic Wars English ships showed a marked reluctance to surrender whereas the Spaniards seem to have regarded striking as an almost automatic course of action after a certain amount of damage had been suffered. These national characteristics could be simulated by adding 1 to the dice of English ships testing for surrender and deducting 1 from the dice of Spaniards.

2 hull boxes destroyed	—	1	=	strike
3 hull boxes destroyed	—	1,2	=	strike
4 hull boxes destroyed	—	1,2,3,4	=	strike
5 hull boxes destroyed	—	1,2,3,4,5	=	strike

When a vessel has had all five hull boxes crossed through but has still not struck it will test for surrender every move that it is fired into.

When a ship receives the surrender of another it must heave-to for one move and send a prize crew across in a boat (boats move at 4 in per move). The prize crew will take four moves to take over the ship and may then sail her away, but may not use her to fight.

Signalling

The introduction of a proper system of signalling at the end of the 18th century almost revolutionised naval operations, but communication by flag hoist was still a fairly slow and clumsy business and so it is a good thing to include in our rules a restriction upon orders issued from the Flagship so as to inhibit the unrealistic telepathic flexibility which often characterises the manoeuvring of wargames fleets.

The signalling and communication rules given in this section are intended to be nothing if not a limit upon the wargamer's freedom of action, so that they have inherent within them a 'controversy factor'. For this reason they are best used in wargames where an umpire is present so that obsolete orders, misleading signals and wrong interpretations can be carried out under his despotic insistence reather than leaving it to tender consciences. Two honest players, however, can still make fair use of the rules.

Most Admirals took the opportunity before battle to explain their tactical ideas to their Captains in a council of war. In our wargame this council of war takes the form of a written plan of action, including the orders and roles of each ship or squadron, which is handed to the umpire, if present. In the absence of any signal to the contrary all ships will do their best to carry out this plan.

In addition to his written council of war orders each player writes a 'signal book' consisting of as many orders as he can think of, with each order allocated a number. Each ship is also allocated a number. If it is required that, say, *Dreadnought, Montague,* and *Russel* should make all sail and attack the enemy van, the Admiral will call out the appropriate numbers, perhaps 6, 7, 10 (the ships' numbers), 17 (the order to make all sail), 14 (the order to attack the enemy van). The umpire (or perhaps, in the absence of an umpire, the player's sense of fair play) will note the order and ensure that it is carried out strictly.

Only six flags may be hoisted (ie, numbers called) per move. If an order is to be given that is not included in the signal book it may be given word by word, but each word is considered to be one flag.

Signals can only be given if the signalling ship has a mast left to hoist them from and the mast must be in direct line of sight of the ship or ships which are being spoken to. A frigate placed behind the line of battle can be used to repeat signals.

Each signal is hoisted (ie, announced) at the beginning of the move and then is assumed to be read and acknowledged during the move so as to become effective at the beginning of the next. Thus there is always at least one move's delay between an order being given and being acted upon, perhaps more if more than six flags are needed. If a repeat ship is necessary to pass on the order it will repeat on the move after the Flagship has signalled, and the ship spoken to will act on the move after that, (ie, a two-move delay).

Sails and anchors

A ship may not anchor in a greater depth of water than 50 fathoms. It takes one move to drop anchor and ten moves to weigh. A 2 in piece of thread should be let out from the bows of the model to represent the cable, around which the ship will swing, being always downwind or down-current. It takes two moves to set or take in sail.

Shoals and shallows

When these feature in the fighting area they can be indicated in a number of ways:

1 By marking the playing area with sea-bed contours of chalk or string up to the five-fathom line.
2 By giving each player and the umpire a map or chart showing shallows.
3 If one player is defending his own shore line he and the umpire may be given an accurate chart, but his opponent only a general indication as to the whereabouts of shallows.

When a player has no chart he may sail across suspected areas at a maximum speed of 2 in per move and take soundings with a lead-line. When this is happening the umpire, or the other player with the chart, will call out the depth, say, 'five fathoms and shallowing', so that the sounding ship need not run aground.

The depths of water in which ships will run aground are as follows:

Ships of the line	—	3 fathoms
Frigates	—	2 fathoms
Brigs, sloops, etc	—	1½ fathoms
Cutters	—	1 fathom

The shock of running aground may spring the masts, throw one dice per mast:

Speed 5 to 8 in per move	—	1	= mast falls
Speed more than 8 in per move	—	1,2,3	= mast falls

A ship which runs aground at high tide is to all intents and purposes lost. One which runs ashore at low tide may be towed off at high tide, and one which runs aground in non-tidal waters may be towed off if a 5 or 6 is thrown on a dice.

Ships striking rocks sink!

Towing

A dismasted ship or one which is aground may be towed by another ship, or by the ship's boats. A 4 in length of thread is used to represent the towing rope and it takes six moves to pass from one ship to the other. Ships' boats may be launched in one move but they may not operate in gale force winds.

Ships move at half speed when towing, and ships' boats at 1 in per move. A grounded ship being towed off the shallows will only move 1 in per move, or ½ in per move if towed by boats, until it is outside the five-fathom line.

Ship data

There was no standardisation of ship design and armament in the 18th century, but the details given below can be taken as being fairly representative of the period when calculating firepower and boarding factors.

Rate	Guns	Gun decks	Crew	32 pdr	24 pdr	18 pdr	12 pdr	9 pdr	6 pdr	Carronades
1st	120	3	900	32	34		34	16		6 × 24 pdr
1st	110	3	900	30	32		32	16		4 × 24 pdr
1st	100	3	850	28	28		44			2 × 32 pdr
2nd	98	3	750	28		30	40			6 × 18 pdr
3rd	80	2	700	30	32			18		4 × 24 pdr
3rd	74	2	650	28	30			16		2 × 32 pdr
3rd	64	2	550		26	26		12		
4th	50	1	350		22		22		6	6 × 24 pdr
5th	40	1	320			28		12		
5th	38	1	270			28	2	8		
5th	36	1	250			26	2	8		
5th	32	1	220				26		6	
6th	28	1	200					24	4	
6th	20	1	160					20		

Chapter 7

The Battle of Ushant, 1795

In 1795 the revolutionary government of France had decided upon an invasion of Ireland. As a first step to invasion they ordered a concentration of their fleet on Brest and, in accordance with these instructions, Citizen Admiral Paul Egalité d'Hanley (late bo'sun in the old Royal Navy) had taken his fleet out of Cherbourg and sailed down the Channel with a following wind. Fortunately there was a thick sea mist when he left the port so he had little difficulty evading the English blockading squadron; with luck he would be able to do the same when approaching Brest.

His luck did not hold out. As he rounded Ushant his scouting frigate, *Le Thémis,* reported enemy sail in sight in the south-east. These were shortly afterwards identified as seven English sail of the line and a frigate, part of the Channel Fleet, under Rear-Admiral Sir Horatio Hague. A fight was inevitable as they sat right between the French Fleet and their destination, so d'Hanley ordered his ships cleared for action.

He had the following ships under his command:

Le Républicain (Flagship)	110 guns
L'Intrépide	110 guns
Le Tyrannicide	74 guns
Le Scipion	74 guns
L'Achille	74 guns
Le Vengeur	74 guns
Le Thémis (frigate)	40 guns

Sir Horatio had a force of roughly equal strength:

Britannia (Flagship)	100 guns
Dreadnought	98 guns
Majestic	74 guns
Albion	74 guns
Venerable	74 guns
Hannibal	74 guns
Agamemnon	64 guns
Euryalis (frigate)	36 guns

D'Hanley's objective was to arrive at Brest as little damaged as possible. The English would be fought later when the combined French Fleets would sally forth from Brest and overwhelm them with superior force. His instructions to his Captains in the case of meeting an enemy force, therefore, were to stand off and wreck the enemy's rigging with dismantling shot. Any injury which could be inflicted on the enemy once he was crippled (without incurring significant damage) would be done, but the *principal* object was to reach port in good order. All French guns were loaded with chain shot.

Sir Horatio Hague's objective was the complete destruction of the French Fleet. His intention was to break the enemy line about half way and, to allow for a quick reaction to circumstances without the delays of signalling, he would lead the line of battle with his Flagship. All his ships' guns were double-shotted.

The battle opened with the two fleets sailing on directly opposite courses, the French reaching on the port tack and the English reaching on the starboard tack. The wind blew from the north-east. The French Flagship was fourth in the line with *Le Thémis* positioned to starboard as repeat ship.

Moves 1, 2, 3 and 4—With both fleets set on what was apparently a collision course, the Admirals were in a quandary as to how to apply their chosen tactics. Sir Horatio contemplated a turn to windward and then a run down to leeward breaking the enemy line when they drew level. However, when a string of signal flags broke out from *Le Républicain*'s yard arm (ie, the opponent said 'Signal' and wrote his order down) he deferred his change of course. The following move the signal was repeated on *Le Thémis'* halliards and was acknowledged by the rest of the fleet. As the signal became executive the French plan became apparent to Sir Horatio; *Le Tyrannicide* bore away to leeward and fired chain shot at the *Britannia*'s masts and rigging whilst the remainder of the fleet came up to the turning point preparatory to turning in succession.

Le Tyrannicide fired at extreme range. Throws on the average dice were 3 and 5 = 8 − 3 (range) = 5, or $\frac{1}{2}$ × 44 (fire factor). Therefore 22 mast hits had been scored on *Britannia*.

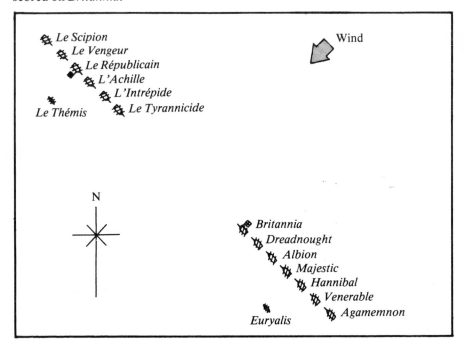

Viewed from behind the French Fleet Le Tyrannicide *bears away and opens fire.*

Move 5—Sir Horatio saw the Frenchman's mistake and seized his opportunity. As they turned to leeward they went on to a broad reach and as they had not reduced sail they increased speed and opened up big gaps in the line. The *Britannia*'s helm went up and she turned to port on a broad reach and bore down on the French line at right angles. The *Dreadnought* and the rest followed in line and would turn in succession as they reached the point. In the French line *L'Intrépide* came up to the turning point, bore away and fired chain shot at the *Dreadnought* as she passed. *Le Tyrannicide* fired raking chain shot into the approaching *Britannia,* who was unable to reply.

1 *Tyrannicide* fired at *Britannia:* 7 (dice) – 1 (medium range) + 1 (raking) = 7. 1 × 44 (fire factor) = 44 mast hits on *Britannia.* **2** *L'Intrépide* fired at *Dreadnought*: 8 (dice) – 3 (extreme range) = 5. $\frac{1}{2}$ × 56 (fire factor) or 28 mast hits.

Move 6—*Britannia* sailed into the gap between *L'Intrépide* and *L'Achille* and she fired a tremendous, double-shotted raking broadside into each as she broke through the line.

1 *Britannia* fired into the stern of *L'Intrépide.* 7 (dice) + 2 (short range, double shot) + 2 (rake stern) = 11. $2\frac{1}{2}$ × 52 (fire factor) = 130! *L'Intrépide* was reduced to four fifths firepower. **2** *Britannia* fired into *L'Achille*'s bows. 8 (dice) + 2 (double shot) + 1 (rake bows) = 11. $2\frac{1}{2}$ × 52 (fire factor) = 130! 4 on an ordinary dice = 130 × $\frac{1}{4}$ = 32 mast hits: *L'Achille* was also reduced to four fifths firepower. **3** *L'Intrépide* and *L'Achille* fired long-range chain shot at *Albion,* third in the line, and scored 26 and 33 respectively. **4** *Tyrannicide* fired chain shot at extreme range at *Majestic* and scored 11.

Move 7—*Britannia* sailed through the line, followed by *Dreadnought,* and

Britannia *breaks the line.*

turned to starboard. *L'Achille* was compelled to turn to starboard to avoid collision and this manoeuvre put her in position to be raked again by *Britannia,* whilst *Le Républicain* had to put her helm hard over to starboard to prevent herself running aboard *Dreadnought.*

1 *L'Intrépide* fired at extreme range into *Hannibal*'s mast and scored 22 hits. **2** *L'Achille* fired chain shot at *Dreadnought* but at such short range that it had to hit the hull, so only causing half effect and scoring 21 hull hits. **3** *Britannia* raked *L'Achille*'s bows at short range with single shot, scoring 104 hits! The Frenchman was reduced to two fifths firepower. **4** *Dreadnought* fired her starboard, double-shotted guns into *Le Républicain* and scored 96 hits. **5** *Dreadnought*'s starboard broadside fired at *L'Intrépide* at long range and scored 60 hits.

No other French ships were in either position or range to make reply but nevertheless all those still loaded with chain shot fired off their guns and loaded with round shot. They could see the way the battle was going!

Move 8—To d'Hanley's chagrin neither *Le Tyrannicide* and *L'Intrépide,* nor his repeat ship *Le Thémis* had a clear view of his Flagship so he was unable to signal to his two leading ships. Nevertheless he ventured to hope that their Captains would have the common sense to return to the fight on their own initiative. His faith in them was vindicated, for straight away they turned about and beat back to the mêlée (ie, it was agreed between the players that a dice throw of 4, 5 or 6 per ship would indicate proper initiative in the Captains. The dice were 4 and 6).

Albion and *Majestic,* abandoning the strict order of the line ahead, sailed into the French line whilst *Britannia* continued her turn to starboard and *Dreadnought* sailed on and laid herself alongside the much punished *L'Achille.* The coming of *Albion* and *Majestic* into the gap almost together thwarted the French Flagship's attempt to break through to the other side, and caused great congestion in the rear with *Le Vengeur* and *Le Scipion* being compelled to turn to port and starboard respectively to avoid collision.

1 *L'Achille* fired round shot into *Dreadnought*. She scored 27 hull hits and 13 mast damage points. **2** *Dreadnought* hit *L'Achille* scoring 52 hull hits and 39 at the base of her masts which brought her foremast down. **3** *Le Vengeur* fired into *Britannia* as they passed on opposite courses. She scored 88 hull hits and 44 mast hits which brought the Flagship's mast down. **4** *Britannia* put 65 shot into *Le Vengeur*'s hull. **5** *Albion* raked *Le Scipion* with double shot. 88 hull hits. **6** *Le Républicain* raked *Albion*'s stern—84 points. *Albion* reduced to four fifths firepower. **7** *Majestic* fired double-shotted into *Le Républicain,* scored 66 hull hits and reduced her to four fifths firepower.

Move 9—*Britannia* was brought to a halt by the drag in the water of her fallen mast and so was unable to prevent *Le Vengeur* coming under her stern and pouring fire into her transom. As *Le Vengeur* came round, though, she fell aboard the *Albion* and the two ships locked together.

L'Intrépide and *Le Tyrannicide* were slowly beating their way back to the fray, but the French centre and rear were completely disorganised.

1 *Le Vengeur* and *Albion* fired 44 and 70 hulling hits into each other and were both reduced to four fifths firepower. **2** *Le Vengeur* fired her starboard broadside into *Britannia*'s stern causing 44 hull and 33 mast hits. **3** *Dreadnought* and *L'Achille* fought out their duel. *Dreadnought* suffered 36 hull hits and inflicted 72 and 54 mast hits. *L'Achille*'s main mast fell, she was reduced to one fifth firepower and *Dreadnought* called upon her to surrender. A dice was thrown—1. *L'Achille* struck her flag. **4** *Majestic* and *Le Scipion* fired at each other in passing. *Le Scipion* scored 53 hull and 37 mast hits on *Majestic* and received a mere 11 hull and five mast hits in reply. *Majestic* also fired into the approaching *L'Intrépide* at long range and scored 22 hull hits. **5** *Le Républicain* came under fire from the three rear ships of the British line, *Hannibal, Venerable* and *Agamemnon,* and suffered a massive 188 hull hits and 126 mast hits which reduced her to two fifths firepower and brought a mast down. She replied with the four fifths effectiveness she had at the beginning of the move,

The French van is isolated and the centre thrown into confusion.

On the left of the photograph the French Flagship is set upon by the three ships of the English rear. L'Intrépide *and* Le Tyrannicide *are crawling close-hauled back into the fray.*

but only scored 22 hits on *Hannibal* and 44 on *Agamemnon*.

Move 10—*Dreadnought* hove-to and sent a prize crew in a boat over to *L'Achille* to take possession.

The wreckage of its fallen mast halted d'Hanley's Flagship and so she was unable to escape from the remorseless fire of *Hannibal* and *Agamemnon*. *Le Scipion* attempted to come to her assistance and sailed under *Hannibal*'s stern raking her, but she was attacked in her turn by the *Venerable*.

Britannia was now sailing again having cut away her wrecked mast. She brought her head across the wind and steered alongside *Le Vengeur* who was locked in combat with *Albion* and attempting to board.

L'Intrépide and *Le Tyrannicide* slowly sailed close hauled up to the fight but were not able to engage during this move. *Majestic* sailed out and fired into them.

1 *Albion* fired 44 hull and 33 mast hits into *Le Vengeur* reducing her to three fifths firepower. *Britannia* also fired at her and hit her with 21 hull and 15 mast hits. **2** *Le Vengeur* fired 26 hull and 13 mast shots into *Albion* and 18 and five into *Britannia*. *Le Vengeur*'s boarding party swarmed on to *Albion*'s decks, but being only 96 against 128 they were repulsed by the *Albion,* who counter-boarded. **3** *Le Scipion* scored 44 hull hits on *Hannibal* but suffered 55 hits from *Venerable* who was positioned on her bow. **4** *Le Républicain* fired half her port broadside (all that would bear) at *Agamemnon* and scored eight. She would have been able to fire half her starboard guns into *Venerable*'s stern but the guns were masked by the wreckage of her foremast. **5** *Hannibal* and *Agamemnon* hit the Flagship with a total of 63 hull and 39 mast shots. Her main mast fell. **6** *Majestic* fired into *L'Intrépide* at medium range and scored 44 hull and 33 mast hits, reducing her to three fifths firepower. **7** *L'Intrépide* fired back at *Majestic* and scored 55 hits in the hull and 33 on the mast bases. She was reduced in firepower by one fifth and her foremast fell.

Move 11—*Hannibal* and *Agamemnon* sailing on either side of *Le Républicain* were compelled to turn sharply to port and starboard to avoid collision and so were forced to leave the Flagship in peace this move. She was unable to escape though, because of the drag of the newly fallen main mast in the water.

The *Majestic*'s fallen mast also brought her to a halt and *L'Intrépide* took the opportunity to steer under her stern and rake her. *Venerable* came up behind *L'Intrépide* and to avoid her *Le Tyrannicide* had to turn starboard, which brought her head to wind and she was taken aback.

Dreadnought had left her boat and prize crew rowing across to *L'Achille* and was now sailing down on the port tack to support the *Majestic,* whilst *Britannia* sailed slowly on at reduced speed to engage *L'Intrépide.*

1 *Le Vengeur* fired into *Albion* and scored 26 hull and six mast hits. **2** *Albion* fired at *Le Vengeur* at point-blank range and the *Hannibal* fired at medium range. She suffered 81 hull hits and 54 mast hits and was reduced to two fifths firepower whilst her foremast fell down across *Albion*'s fo'c's'le. The cutlass-swinging *Albion* crew poured on to her and carried all before them. The ship was taken. **3** *L'Intrépide* raked the crippled *Majestic*'s bows and scored 34 hits. **4** *Britannia* scored 32 hull and 16 mast hits on *L'Intrépide* at short range and 21 and five into *Le Républicain*'s stern at long range. **5** *Venerable* raked *L'Intrépide*'s stern and scored a smashing 88 hull and 22 mast hits. She fired double shot from her hitherto unengaged port broadside at *Le Tyrannicide* and scored another 88 and 22 on her.

Moves 12, 13 and 14—*Agamemnon* and *Hannibal* turned and steered for the crippled *Le Républicain.* One broadside from *Agamemnon* and the Flagship surrendered. D'Hanley became a prisoner.

L'Intrépide was unable to extricate herself from a very one-sided combat with *Dreadnought, Majestic* and *Britannia* and eventually surrendered when reduced to one fifth firepower.

The remaining French men-of-war, *Le Tyrannicide, Le Scipion* and *Le Thémis* crowded on sail and fled on course for Brest as soon as the Flagship was seen to strike her flag. Sir Horatio hoisted the signal 'general chase' but by the time his frigate and three battleships with undamaged rigging set off in pursuit the French were well away and could not be caught. Sir Horatio was not too troubled about that. Four ships of the line and an Admiral captured was not a bad bag.

Victory made Hague feel magnanimous. Although he could not regard a Jacobin Admiral as being quite a gentleman he nevertheless entertained d'Hanley right royally in his stateroom aboard the *Britannia.* He discussed, perhaps just a little didactically, the excellence of close-range, double-shotted tactics and compared them with the stand-off-and-cripple method, to the disadvantage of the latter; he deprecated the taking of the leeward position and roundly condemned the folly of allowing gaps to open in the line, especially when the opposing Admiral was a valiant, aggressive, brilliant, talented bulldog whose military genius was world renowned. The tone was patronising but d'Hanley said nothing. He was safer in the *Britannia*'s stern cabin than he would have been in Brest where the guillotine would be his richly deserved reward for his performance. He sipped at his port and let Hague blabber on.

Chapter 8

The Ironclad period, 1865–1885

In the early 1860s the world's navies started to look critically at their battle-fleets. Technology, in the form of explosive shells, had at last overtaken the stately but inflammable old 'wooden walls' which had been the mainstay of navies since the 16th century. It was necessary to keep shells out of the hull; armour was needed; battleships should be Ironclads.

This much was agreed; what was not generally agreed was the form which Ironclads should take. Conservatism suggested a gun armament in broad-side, but ram enthusiasts demanded cut away sides to give fire support to the battle-winning ramming attack, whilst the real progressives deplored the encumbrance of guns at all in what should simply be a steam ram. The gun school, when it was not squabbling over the relative merits of turrets, barbettes and batteries, was split into the monster-gun 'one devastating shot every five minutes' faction and the medium gun 'faster fire equals faster hitting' party; concurrent with these arguments the advocates of long range and habitability were proclaiming the excellence of masts and yards and decrying the unseaworthy turreted monitors so beloved by their opponents, whilst both 'turretists' and 'mast-ers' came together in a condemnation of the 'circular hulls for steady gun platforms' heresy.

All concepts took material form in at least one ship and so until the late 1880s, when the fittest species at last started to emerge, the world's battlefleets were collections of often unique samples. No fleet actions took place during this period to vindicate or disprove designs so there is no real way of knowing which ship types were practicable and useful and which were just plain daft. Nor does history tell us how these heterogeneous collections of warships, with all their incompatible characteristics, were to be employed in the mutual co-operation of battle.

The main features of interest of this period, then, lie not in the strategic movements of large fleets but in the designs and fighting qualities of the individual ships and in the tactics necessary to allow the mixed bag components of a squadron to fight in support of each other. To highlight the characteristics of the different ships my rules for this period are much more detailed and 'analytic' than the other sets described in this book; we operate on a short cycle time of one minute to a game move, account separately for the loading times of guns, fire them individually and plot the point of impact, penetration and damage effect of each shell. Of course, no wargames rules, however ingenious, can properly evaluate the true worth of a warship's design because no real account can be taken of a ship's fundamental qualities such as stability, roll characteristics, construction, etc, which have a great bearing on its ability to inflict and suffer damage. A suitable set of rules, though, will allow a ship's strictly military properties, gun layout and armour distribution, speed and manoeuvreability, to be effectively simulated so that the results of combat between ships of different types will accurately reflect their strengths and weaknesses in these respects.

It naturally follows that by increasing the details and complexities of the rules we are slowing down the playing time and imposing more paperwork on the players. About half a dozen ships is the most that a single player can reasonably handle at one time using these rules but with the great variety of ship types available in an Ironclad fleet a six-ship squadron gives a very good wargame.

Weapons

The weapons of the Ironclad battleship were in descending order of importance the gun, the ram and the torpedo. One or two ships were designed purely and simply as steam rams in which no guns were carried, but most were gun armed, with a strengthened spur for ramming, on the belt and braces principle. Towards the end of our period most warships were also equipped with locomotive torpedoes and launching gear.

Guns

At the beginning of the Ironclad period the weight of iron armour precluded its use on more than one gun deck so that the guns carried were less numerous but bigger than those mounted in the 'wooden walls'. By later standards, however, there were a lot of them, they were relatively small and the ships were, in effect, single-decked frigates. In the Royal Navy the guns were initially breech-loaders but a dissatisfaction with this type led to a reversion to muzzle loading which lasted throughout our period.

Increases in armour thickness led to an increase in the size and penetrative power of guns and, because thicker armour could only be employed if a lesser area of the ship's side was covered, the guns, which were invariably behind armour, became fewer. Thicker and stronger armour inevitably resulted in bigger and bigger guns until by the early 1880s guns of up to 17.7 in calibre were afloat. These guns being muzzle-loaders, they needed hydraulic rammers to load them but they were really too big for the technology of the time and their firing rate of around one shot in five minutes meant that they were unlikely to hit the target (because the time period between shots prevented effective correction). Moderate-sized guns of between 10 and 12 in were probably more effective.

Whilst the technology of gun construction was improving to the extent that these enormously powerful weapons could be built and mounted, however, aiming and firing techniques had hardly altered since the 17th century. They were still laid by the gun Captain standing well back beyond recoil distance and lining up the target with a bead foresight and V rearsight, so it is hardly surprising that effective ranges were short, about 2,000 yards at the most, and accuracy low.

Guns were mounted either in batteries, turrets or barbettes.

Turrets were usually of the Ericson type, which was mounted on the upper deck and rotated by a central spindle, or of the Coles type, which was mounted on the main deck and had its upper part with gun ports protruding through the upper deck and was rotated by pinions around the circumference of its base. There were variations on the theme, however, perhaps the oddest being the fixed turret on HMS *Hotspur* within which was a turntable with a single gun which turned inside the turret and fired through ports around its side.

Where the guns were hydraulically loaded the turret would have to be turned between shots to align the barrels with the loading tubes, but this did not necessarily slow the rate of fire as with manually-loaded guns the turrets were turned away from the enemy to protect the vulnerable ports. For these reasons it was usual for twin guns to a turret to be fired together rather than at equal intervals.

A battery was simply the age old system of mounting the guns in broadside, although in the later box-battery ships cut-away sides and angled ports allowed a degree of axial fire. Because the armament had to be duplicated, smaller guns were usually mounted in batteries than were used in turrets, although there were instances of 11 and 12 in guns being used in broadside.

Barbettes were fixed, roofless cylinders of armour containing turntable-mounted guns which fired over the top. The advantage of the barbette system over turrets was that the guns were carried higher so that they could be worked in weather which would put lower-mounted guns awash and out of action. It was also thought that barbette guns would be able to fire down on the unprotected decks of an enemy alongside, although this seems rather fanciful. More accurate aiming was also claimed for barbette guns, due to the crews being in the open and not having to peer through narrow ports, but to be set against this was the danger to the partially exposed crew of machine-gunfire. In a couple of instances the problem of crew exposure was solved by employing disappearing mountings on which the gun would fire over the barbette parapet but swing below and behind it for loading.

Rams

The invulnerability of the early Ironclads to the smoothbore gunfire of the time, combined with the new found freedom of mobility afforded by steam power, led to the ram gaining a great, though unwarranted, importance in naval tactical doctrine during the 1870s and '80s. In an age of very close fighting it was a useful second string weapon, which cost very little to equip ships with, but its battle-winning power became so exaggerated that battles with ships in line abreast were envisaged and gun layouts were made to conform to this requirement for end-on fire. In some ships the guns were left out altogether.

In practice, however, it was all too easy to avoid the ram so long as the target ship had steerage-way on and the only ships likely to succumb to a lethal stab from an enemy ram were those which had already been crippled by gunfire. A great many peace-time collisions occurred involving ram-bowed ships and the results of these accidents do show that a penetration by a ram was likely to be fatal.

Torpedoes

The term 'torpedo' was used for all underwater weapons, including mines. Non-static torpedoes fell into three categories; the Harvey, or towing torpedo, the Spar torpedo and the Whitehead or Locomotive torpedo.

The Harvey was an explosive charge towed behind a small steam boat and, it was hoped, brought into contact with an enemy hull. The casing was designed rather like a modern paravane so that the torpedo ran out to one side of the towing boat.

The Spar was an alarming contraption consisting of an explosive charge at the end of a long pole mounted in the bows of a small boat. The charge was rammed into the side of an enemy ship and ignited electrically; with luck it was the enemy, not the torpedo boat, which was sunk.

The Whitehead was an early form of the modern torpedo, which could be launched from either small torpedo boats or battleships. They were rather crude at this time and were limited to a range of about 800 yards at 12 knots and as gyroscopic control had not yet been developed for them they tended to follow an eccentric course which made their use something of an art rather than a science.

Torpedoes were essentially weapons for use by small boats against ships at anchor and so their main employment was in night attacks. Such torpedo boats as accompanied the fleet were carried aboard battleships and launched from davits when required. Guard boats and Nordenfeldt and Gatling machine-guns were the principal defence against torpedo attacks.

Ship types

Belt and battery ship, circa 1865

The guns are mounted in broadside but on one deck only. The low power of guns means that thin armour, $4\frac{1}{2}$ in approx, is sufficient, so the whole of the ship can be protected.

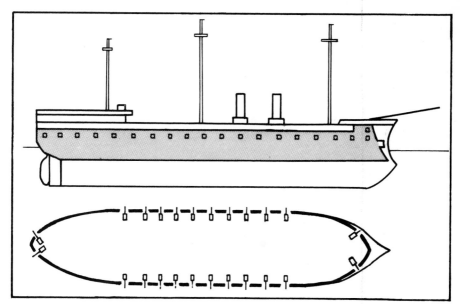

Box-battery ship, circa 1875

Bigger and more powerful guns mean fewer of them and thicker armour. Armour is limited to a waterline belt and a box amidships in which the guns, magazines and engines are contained. The ports are angled and the ship's sides cut away to allow ahead and astern fire.

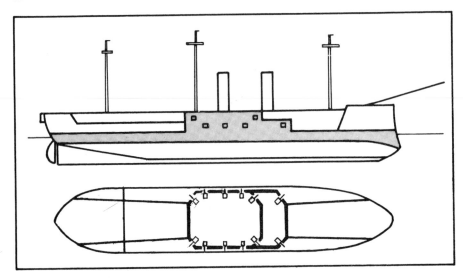

Breastwork monitor, circa 1872

A mastless ship to allow unobstructed fire from the turrets. In order to cover the whole hull with thick armour she is built with a low freeboard, with the turrets mounted on an armoured breastwork to give them sufficient height above the water. Poor living conditions make them unsuitable for blockade work or tropical conditions, so less powerful masted ships continue to be built.

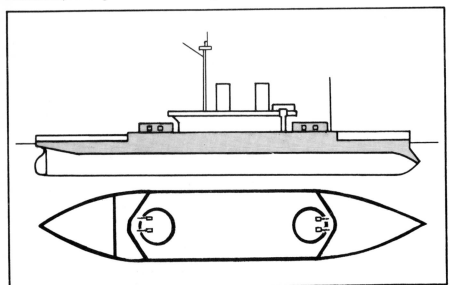

Central citadel ship, circa 1880

The thickest armour afloat is carried but is limited by its weight to a citadel amidships containing engines and magazines. The shortness of the citadel forces an echelon arrangement of the turrets. The unarmoured ends are protected by an armoured deck and cork-filled cells.

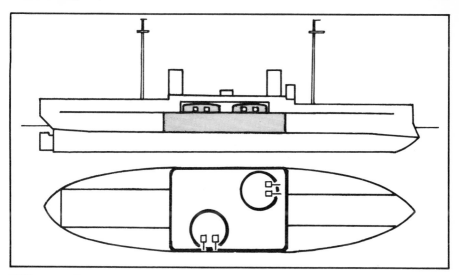

The rules

Scale

Gun ranges in this period are still quite short so I use a scale of 1:1,200 for models and for distances. In this scale 100 ft becomes 1 in and 1 knot becomes 1 in per move, each move representing one minute of real time.

Movement

Under steam—Ships will move 1 in per knot of speed up to their maximum. Speed may be increased by up to 2 knots per knot of speed and speed reduction is at a similar rate, except where it is enforced by sinkage or bow damage, when the speed will be reduced to the new maximum immediately.

Under sail—The additional displacement of armour plating caused Ironclads to be very poor sailors compared with their wooden predecessors but on most ships masts and yards were retained for economical cruising. In action top masts and yards were sent down and the ships proceeded under steam alone, but as sails can be useful for removing crippled ships from action a simple set of sailing rules is given here.

If a ship's best speed under sail is not known we use a figure based on its maximum steaming speed:

Single hoisting screw	—	Full steam speed
Single non-hoisting screw	—	$\frac{3}{4}$ steam speed
Twin screws	—	$\frac{1}{2}$ steam speed

Speed depends upon the point of sailing and wind strength:

	Beating	**Reaching**	**Running**
Light Wind	$\times \frac{1}{4}$	$\times \frac{1}{2}$	$\times \frac{1}{3}$
Fresh wind	$\times \frac{1}{3}$	Full	$\times \frac{2}{3}$
Gale	$\times \frac{1}{4}$	$\times \frac{3}{4}$	$\times \frac{1}{2}$

Loss of speed under sail is pro rata with the destruction of masts, so that a three-master losing one mast will be reduced to two thirds speed. Speed under sail will not, of course, be affected by damage to funnels and engines. Reductions due to bow damage and sinkage will be applied at the full rate.

Top masts and yards may be sent up and set, or sent down in ten minutes (ten moves). Sails may be taken in or set in five minutes (moves).

Turning

Where a ship's turning diameter is known it will turn on a card circle, marked around the circumference in inches, of that size on a scale of 1 in to 100 ft. Otherwise we use the following circles:

Ships over 400 ft long	—	500-yard turning circle (15 in)
Ships over 300 ft long	—	400-yard turning circle (12 in)
Ships over 200 ft long	—	300-yard turning circle (9 in)
Ships over 100 ft long	—	200-yard turning circle (6 in)
Ships over 50 ft long	—	100-yard turning circle (3 in)
Less than 50 ft long	—	50-yard turning circle ($1\frac{1}{2}$ in)

Ships classified as rams will be exceptionally handy and so will turn on a circle the next size smaller for their length. Ships under sail alone will turn on a circle one size larger, and twin-screw ships with rudders out of action will turn on circles two sizes larger.

Firing

Many of the big guns of this period will not be able to fire every move, so it is necessary to keep a record of the moves on which a gun is able to fire and those on which it is being loaded. The gunnery record chart consists of horizontal lines representing moves and vertical columns representing guns. When a gun is fired the fact is noted in its column and the appropriate number of loading moves are marked on after it. Note is made of the type of projectile being loaded, shell or shot.

The crude sighting equipment of the time meant that no gun, however large, was likely to hit its target at a range greater than a mile. Indeed a slow-firing big gun would be less likely to score a hit than a faster-firing medium gun, because of the difficulty of correcting when there is a long interval between shots, but this would be largely offset by the flatter trajectory of the big gun. For the sake of simplicity we give each gun the same effective range and probability of hitting.

Maximum range is 2,000 yards (60 in). Up to 500 yards (15 in) is short range, 1,000 yards (30 in) is medium range and 2,000 yards is long range. One ordinary dice is thrown per gun firing and hits are scored by throwing a 6 at long range, 5 or 6 at medium range and 4, 5 or 6 at short range. This means that our gunfire is more accurate than it really was, but this is no bad thing as it tends to speed up the game.

Rate of fire

Guns may fire at the following rates. The intervals between firing moves are assumed to be spent in loading; a gun in a twin turret may not load in the same move in which its sister is firing.

7 & 8 in guns	—	1 round per min (move)
9 & 10 in guns	—	1 round per 2 mins
11 & 12 in guns	—	1 round per 3 mins
16 & 17.7 in guns, etc	—	1 round per 5 mins

SHIP — H.M.S Captain

GUNS / MOVE	FORWARD			AFT		
	7 in	12 in	12 in	12 in	12 in	7 in
1	Shell	Shot	Shot	Shot	Shot	Shell
2						
3	Fire	Fire		Fire	Fire	
4	shell-f	Load	Shot	Load	Shot	Shell-f
5	,,	↓	↓	↓	↓	,,
6	,,	Fire		Fire		,,
7	,,	Load	Shot	Load	Shot	,,
8	Shot-f	↓	↓	↓	↓	Shot-f
9	,,	Fire		Fire		,,
10						
11						
12						

A gunnery record card *At the beginning of the battle the 7 in guns were loaded with shell and the 12 in guns with shot. Firing commenced on the third move, with the 7-inchers firing one round of shell every move until Move 8 when they changed to shot. The 12 in guns fired shot at a rate of one round every three minutes.*

Armour penetration

The two types of projectile in use were shot and shell. Shells contained bursting charges of gunpowder but as they were thin-walled they were likely to break up on hitting armour. Solid shot was used for armour penetration but as it relied on kinetic energy for any damage it caused its destruction area was much more limited.

Armour is given a letter and colour code to denote its thickness. This is explained further in the next section, but suffice to say here that 'A' is the thickest and 'H' the thinnest. The table below gives the maximum armour thickness that shells and shot can penetrate.

Armour penetration table

Gun	Short range		Medium range		Long range	
	Shot	Shell	Shot	Shell	Shot	Shell
7 in	E	—	F	—	G	—
8 in	D	H	E	—	F	—
9 in	C	G	D	H	E	—
10 in	B	F	C	G	D	H
11 in	B	F	B	F	C	G
12 in	A	E	B	F	C	G
16 + in	A	E	A	E	B	F

Shot and shell which fail to penetrate are assumed to explode or break up on the outside without causing any damage.

Armour

For most of our period armour consisted simply of wrought-iron plates with a backing of teak. Towards the end of the 1870s a stronger type, called compound armour, with a steel facing was developed. To simplify our penetration table we must reduce all armours to a common value and I use inches of iron for this purpose. Wood backing 12 in thick is equal to 1 in of iron, and 1 in of compound armour is equivalent to $1\frac{1}{2}$ in of iron. A turret, therefore, which is protected by 10 in of compound armour with a backing of 18 in of teak is considered to be covered with $16\frac{1}{2}$ in of iron.

Armour is given a grade according to thickness (or iron equivalent) and each grade has a colour code which denote armour thickness on the damage record cards.

Armour grading

Inches of iron	Grade	Colour
2 and less than 4	H	Sky Blue
4 and less than 6	G	Blue
6 and less than 8	F	Green
8 and less than 10	E	Yellow
10 and less than 12	D	Orange
12 and less than 14	C	Red
14 and less than 18	B	Brown
18 +	A	Black

Damage record cards

In the other sets of wargaming rules given in this book, damage is simulated by giving each ship a numerical value, in one form or another, and reducing their performance and eventually sinking them when the total value of damage reaches certain predetermined points. This system, although nice and easy to use, tends to suppress the differences between ships of different designs and characteristics and so will not do for these rules whose object is to highlight the strengths and weaknesses of various Ironclad types.

Instead of the numerical value with additions or subtractions which are used as damage records for other periods, we employ a cut away profile drawing of the ships of the fleet and damage is inflicted by plotting the point of impact, seeing whether or not the shot or shell penetrates and, if it does, devastating a part of the ship (and the equipment contained therein) appropriate to the projectile.

The advantage of this system is that the conning tower is destroyed and the ship runs out of control, or a turret is jammed and the ship loses firepower, because the turret or conning tower are inadequately protected, or badly sighted or in some other way faultily designed and not because some arbitrarily decided figure is reached. It has its faults, of course. The drawing is two dimensional so that additional floatation and stability of a large beam is unaccounted for; all shells must be assumed to enter the hull at right angles to its length and height, so that plunging and raking fire cannot be simulated; flooding must be unrealistically equated to explosion or splinter zones and shell bursts, however small, will always devastate the whole width of the ship. The system works well enough for our purposes though, and experience shows that all things being equal the ship with the best balance of military characteristics will win.

The damage record card is made by drawing on to graph paper a profile of the ship's hull showing decks, armoured bulkheads, engine rooms, magazines, steering engines and equipment, guns, conning towers, bridges, funnels, mast bases, etc. The thickness of armour over the protected area is shown by colouring it appropriately for its grade (see table on page 86). Armour belts are taken right down to the keel of the ship. The drawing is made on graph paper with 1 in squares broken down into $\frac{1}{10}$ in sub-squares. The hull from keel to just above the deck or superstructure is drawn so as to fit just within the 1 in height of the square, but the longitudinal scale is 1 in to 50 ft. A vertical column of $\frac{1}{10}$ in sub-squares therefore, represents 5 ft of length, and this is our unit of damage measurement.

Here is a damage record card of HMS *Alexandra,* a British box-battery Ironclad of 1875. The small plan view is just to show the arcs of fire of the various guns.

Most battleships of the Ironclad period were between 250 and 300 ft long and so will fit neatly into six horizontal 1 in blocks. This is very convenient as our first

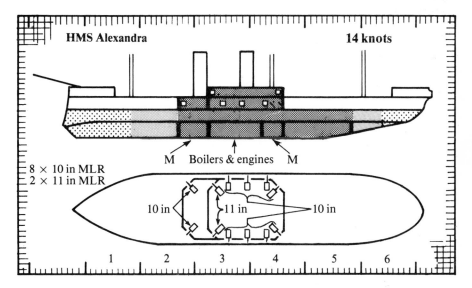

step in locating the point of impact is to throw an ordinary dice to determine the block in which the hit occurred—you will see that the graph blocks of the *Alexandra*'s hull are numbered 1 to 6. If the hull length is very much longer than 300 ft then some other method of locating the box is needed, perhaps by using two dice or decimal dice, or perhaps by drawing playing cards.

We have scored a hit on *Alexandra* and thrown a 6 on the ordinary dice, which indicates that the shell has struck somewhere in the sternmost part of the ship. If the ship is being fired at from ahead or astern, however, only hits on the forward or after half of the hull apply, all others are misses. In this case *Alexandra* is being fired on from abeam so the hit applies; we now throw two decimal dice, one red and one black. The red dice represents the vertical and the black the horizontal co-ordinates in the square so if the scores are 6 and 4 respectively we count up the rows of sub-squares until we come to the sixth and run our finger along until we come to the fourth sub-square in the line and that is the point at which the shell struck. If the hull is armoured at that point the armour penetration table is consulted to see whether or not the shell penetrates.

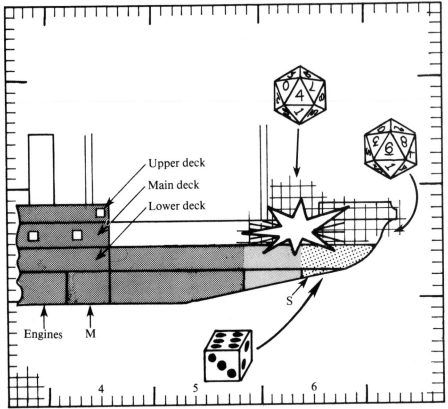

The Alexandra *is hit on the main deck aft. Six sections are destroyed.*

Effect of penetration by shot and shell

On the drawing of the *Alexandra*'s stern above, you will see that around the point of impact six vertical rows of sub-squares, 30 ft of deck, are pencilled in. This is the area of devastation caused by the shell burst or shot splinters. Everything in that area is destroyed or rendered *hors de combat* and if the deck was at or below the waterline that area would also be flooded.

The damage from a penetration, ie, the number of vertical rows blacked out, is equivalent to the destruction value of the shot or shell, as given in the table below.

Destruction value table

Gun	Shell	Shot
7 in	4	2
8 in	6	3
9 in	8	4
10 in	10	5
11 in	14	7
12 in	18	9
16 in	36	18

The following rules apply when blacking out damage:

1 Damage is limited by the decks above and below the point of impact, whether armoured or not.

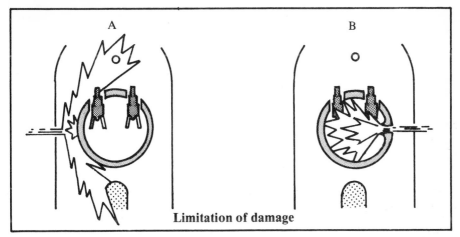

Limitation of damage

On ship A the shot has broken up against the turret armour but the blast of fragments has engulfed the foremast and funnel. On ship B the shot has penetrated but the armour prevents the blast from spreading.

2 Sections are blacked out *equally* on each side of the point of impact.

3 Vertical armoured bulkheads limit damage and armoured units such as conning towers or turrets are immune from non-penetration blast damage.

4 Armour *inside* the hull, eg, turrets and conning towers although not penetrated themselves, may cause shells to explode and shot to fragment; the blast will not affect the armoured unit, but may still destroy unprotected parts of the ship.

5 Magazines are always considered to be protected fore and aft by armoured bulkheads.

The effect of damage on a ship's performance is as follows:

Sinking—When one third of the sections of the ship's hold (the area below the lower deck) is destroyed, the ship will sink by one deck so that the lower deck becomes the new waterline and speed is halved. When half the lower deck sections are destroyed, or two thirds of the hold sections, the ship sinks.

Bows—When a ship's bows are destroyed at its waterline the ship may not travel at more than 6 knots.

Engines and boilers—The ship's speed is reduced proportionately with the number of sections representing engine and boiler room which have been destroyed. A 12-knot ship which has had four of its 16 engine and boiler room sections destroyed will be reduced in speed by a quarter to 9 knots.

Funnels—The loss of all funnels will cause a ship's speed to be reduced by one third, thus a two-funnel ship losing one funnel will reduce speed by a sixth.

Masts—Masts shot down will fall to leeward and mask any guns whose ports are covered. Adjacent ports will also be masked by the ship's rigging, etc. Where upper masts and yards have been sent down the wreckage will take one move to cut away.

Where a mast complete with top masts and yards is shot down it will take three moves to cut away during which period the drag of the wreckage will reduce the ship's speed by half. Any turns away from the wreckage (ie, to windward) will be made on a double-diameter circle.

Steering gear—When the stern-most part of the ship representing the steering gear is completely blacked out by one hit the steering machinery and rudder are destroyed. Twin-screw ships will steer by engines alone on a double-diameter circle. Single-screw ships will have to be taken in tow. When the steering compartment is only partially blacked out, or completely blacked out by an accumulation of damage the steering gear is jammed. Throw an ordinary dice: 1, 2 rudder jammed to port, 3, 4 ahead, and 5, 6 to starboard. Throw one dice every subsequent move, a 6 will indicate that the rudder has been freed.

Conning towers and bridges—Where an armoured conning tower is fitted it is assumed that the Captain (and Admiral if present) are controlling the ship from within it. Otherwise they are on the bridge, or poop.

When the conning tower is penetrated by a shot or shell, or when the bridge or poop are devastated by blast the commanders are rendered *hors de combat* for five minutes (moves). During those five minutes no activity may be commenced on board the ship which would require an order from the conning tower or bridge, eg, change of course, alteration of speed, raising/lowering of masts, change of target to anything but the closest enemy, and so on. If the ship is a Flagship it may not issue orders to the fleet until the five minutes are passed.

Guns—Guns are destroyed when they are engulfed within the blast area of a shot or shell. In broadside ships both the gun on the hit side and its opposite number on the other side are put out of action.

Turrets—A penetration which causes a blacking out of all sections of the turret destroys it. If the blacking out only covers one gun the other is out of action for two minutes whilst the crew recover themselves. A non-penetrating hit on the base of a turret (remember that on a Coles turret this is on the deck below) may cause the turret to jam. Throw a dice:

Armour one grade superior to projectile   —   1, 2, 3 turret jams.
Armour two grades superior to projectile   —   1, 2 turret jams.
Armour three grades superior to projectile   —   1 turret jams.

A 5 or 6 thrown on subsequent moves will free the turret.

Magazines—When a magazine is penetrated, throw a dice to see whether it explodes and destroys the ship:

If a shell hit   —   1, 2, 3 the magazine explodes.
If a shot   —   1, 2   the magazine explodes.

If the magazine is protected by coal bunkers or cork-fitted cells in addition to the armour deduct 1 from the dice. If the magazine does not blow up it is flooded and out of action. Any guns served by it must draw their ammunition from another magazine, slowing their rate of fire by one minute per round. If the ship only has one magazine, each gun is assumed to have three rounds of ready-to-use ammunition, which when exhausted causes the guns to cease fire.

Special protection

Armoured decks—In some ships an armoured deck was installed below the lower deck as a lighter and cheaper alternative to a waterline belt. The purpose of the armoured deck was to seal off the area beneath it, and the machinery contained there, from the effects of a shell burst and thus retain buoyancy.

On the profile drawing of a ship so equipped we represent an armoured deck by drawing a horizontal line in the bottom section of the ship which divides it into two. Any penetration of the lower section will have its damage limited by this

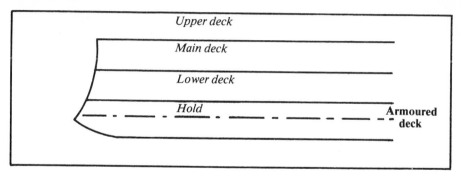

deck, thus halving the effect of shell bursts.

Cork-filled cells—Cork-filled cells around the ship's waterline were used on some citadel ships as lightweight protection for their unarmoured ends. The theory was that the cork would absorb much of the blast of an explosion and partially seal the hole. The theory was a controversial one though, as many pundits claimed that a shell burst would simply scatter the cork and do its full quota of damage. As the efficacy of this type of protection was never tested in action our rules compromise between the two arguments. Hits by 7 and 8 in shells will only cause half damage to cork-protected zones, 9 and 10 in shells will cause three quarters damage and 11 in shells and bigger will scatter the cork and cause full damage.

Disappearing guns—When the barbette is penetrated the gun will naturally be put out of action, but spreading blast from a non-penetrating hit will only put the gun out of action in a move in which it is firing. In a loading or non-firing move the gun is protected by nestling within the barbette.

Machine-guns

The short-range tactics envisaged for Ironclad fleets led to the installation of Nordenfeldt and Maxim machine-guns on hurricane decks and in fighting tops. The role of the machine-guns was anti-personnel, but as we do not concern ourselves with the activities of the ship's crew we interpret the effect of machine-gun fire as a harassment of the functions of the target ship.

1 Maximum range is 250 yards (15 in), medium range is 170 yards (10 in) and short range is 85 yards (5 in).
2 Any machine-gun fired at by another machine-gun at short range is unable to fire that move. Two machine-guns neutralise one at medium range and three neutralise one at long range.
3 Machine-gun fire may cause the crew of an exposed gun to rush for cover and thus miss a move's loading or firing. Throw one dice for the whole group of machine-guns, adding 1 to the dice for each gun additional to one and deducting 2 from the dice if firing is at long range and 1 if it is at medium range; a score of 5 or 6 will neutralise the gun's crew for that move. The crews of disappearing guns within barbettes are immune from machine-gun fire, but gun crews within ordinary barbettes may still be harassed; treat as exposed guns but deduct 1 from the dice for the partial protection of the armour.
4 Against torpedo boats machine-guns are fired as described above. If 5 is scored the torpedo boat is driven off and must retire out of range before

attempting to attack again, a score of 6 will cause the boat to break off action completely.

5   A score of 5 or more against a ship's commanders on a bridge or poop (ie, not in an armoured conning tower) will neutralise them for that move. No action which would require an order from the Captain may be initiated that move and if the ship is a Flagship the Admiral may not issue orders.

Ramming

A ramming attack only takes effect if the blow is made at an angle of between 90 and 45 degrees. At more acute angles the two ships merely bump and scrape sides. The two ships are phase moved until the rammer strikes the rammed. The place of contact is noted and located on the damage record card.

Damage is effected in the same way as gunfire, but it occurs on all deck levels and is inflicted at the rate of one five-foot section on each side of the impact point

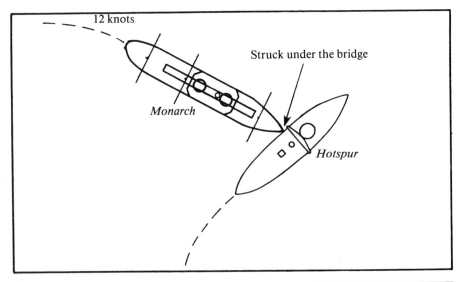

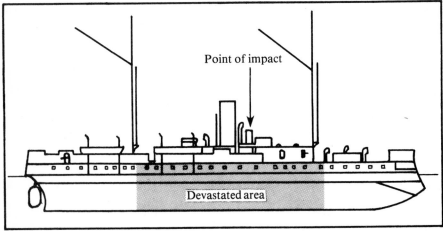

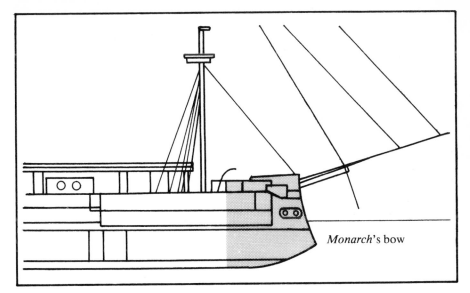

Monarch's bow

Left *The two ships are phase moved into contact. The point of contact is noted.*

Below left *At 12 knots the* Monarch*'s ram has devastated 12 sections before and abaft the point of impact which is sufficient to sink* Hotspur . . . *but* Monarch **(above)** *has an unsupported ram so her bows are damaged for four sections.*

for each knot of speed of the ramming ship. Armour gives no protection to a rammed ship.

If the ramming ship is designated as a 'ram', or if its ram bow is supported by armour it will not suffer any damage from ramming. If, however, it has a 'soft' spur its bows will be damaged at the rate of one section per 3 knots of speed.

As an example we will have *Monarch* ram *Hotspur.*

Torpedoes

Torpedo boats at this period were simply small steam launches equipped with whatever gear was needed for the type of torpedo used. They would be difficult to hit, so we deduct 1 from the firing dice of any big gun aiming at one but being so small a single hit from a shell of any size will sink one without more ado.

Spar torpedoes—It takes five moves not under fire and not travelling faster than 6 knots to fit a spar torpedo to the bow of its boat. The torpedo boat must, in effect, ram its target for the torpedo to strike home. Note the point of contact and throw a dice for effect.

Harvey torpedoes—A length of cotton or piece of thin wire $1\frac{1}{2}$ in long is trailed from behind the towing boat at 45 degrees to its course. The free end of the cable must be towed into contact with an enemy ship. Note the point of contact and throw a dice for effect.

Whitehead torpedoes—Locomotive torpedoes are launched by boats in the direction they are travelling or from tubes or carriages on bigger ships at right angles to the course. They may not be launched from tubes or carriages when the firing ship is travelling faster than 5 knots. The torpedo is represented by a marker, say a matchstick, which may be launched at any point in the move and runs for $1\frac{1}{2}$

moves at 12 knots or, in other words, a total of 18 in before sinking. The torpedo and its target are phase moved, and if a hit is scored the point of contact is noted and a dice thrown for effect.

Effect of torpedo hits

The point of contact, in the lowermost part of the hull, is located on the damage record card. Torpedoes are always assumed to strike below the armoured belt.

Torpedo hits are treated as below the waterline shell bursts, but as they were very crude in design and construction we give them an opportunity to go wrong.

Dice score	Effect
1	— Fails to explode.
2	— Torpedo causes 14 sections of damage. If it is a Spar torpedo its boat is destroyed in the explosion.
3	— 18 sections of damage.
4	— 22 sections of damage.
5	— 24 sections of damage.
6	— 30 sections of damage.

Chapter 9

The Battle of the Hellespont, 1881

In the summer of 1881 all was not well on the North West Frontier. The Russians were massing army corps on the Afghanistan border and although they had stated that they were there in preparation for the autumn manoeuvres, HM Government was dubious; relations had been strained for some time and an invasion of India was suspected. In consequence the Yeomanry and Militia were assembled at their depots and the regiments of the regular army put aboard ship for conveyance to India to reinforce the garrison there.

The route the troopships were to take was Gibraltar, the Suez Canal and thence to Bombay. The Mediterranean Fleet, therefore, was the key force in preventing naval units of the Russian Black Sea Fleet cutting this vital artery, and preparations had been made to convoy all troopers to Alexandria. The Government, however, felt that something more should be done; they wanted a flag-showing force at Constantinople to encourage a Turkish refusal to allow the Russian Fleet passage through their waters. So, at Foreign Office insistence, the Admiralty assembled a Particular Service Squadron at Spithead and to command it they hauled in that gallant old nonagenarian, Vice-Admiral Sir Trafalgar Hague, from his half-pay retreat at Cheltenham.

Sir Trafalgar hoisted his flag in the *Alexandra* and sailed for the Mediterranean immediately his essential supply of port and madeira was safely stowed away. The journey out was made at a steady 5 knots and was without incident. The twenty-third day out of Portsmouth found him nearing his destination, steaming on a northerly course up the Aegean about 20 miles south of Cape Helles. At mid-morning, though, a cry was heard from the lookout in the Flagship's fore-top. Smoke had been sighted. Shortly afterwards the smoke was identified as coming from a squadron of Russian warships emerging from the Dardanelles. For Russian men-of-war to leave the Black Sea through those waters was a breach of treaty and under the circumstances could only be regarded as an act of war. Sir Trafalgar ordered his Squadron to clear for action.

The Particular Service Squadron consisted of four Ironclads:

1 HMS *Alexandra* (Flagship)—1877—Box-battery ship—15 knots. Protection: Engines, magazine, battery—C. Belt—E and F. Guns: two 11 in and ten 10 in.
2 HMS *Inflexible*—1881—Central citadel turret ship—14½ knots. Protection: Turrets and citadel—A, armoured deck and cork-filled cells for the remainder of the waterline. Guns: four 16 in.
3 HMS *Devastation*—1873—Breastwork monitor turret ship—13 knots.

Protection: Turrets—B. Engines and magazines—C. Remainder of hull—
D and E. Guns: four 12 in.

4 HMS *Hercules*—1868—Box-battery ship—14½ knots. Protection:
Battery—E. Belt—F. Guns—eight 10 in guns, two 9 in, and four 7 in.

The intentions of the Russian Squadron were indeed bellicose. Its commander,
Admiral Prince Hanliovitch, had been especially chosen for his pugnacious
approach to international relations. His orders were to attack British troop and
store ships and cause such disruption that they would be compelled to use the
longer and slower Cape route to Bombay. Prince Hanliovitch had six ships with
which to accomplish his task, all of which were built to British plans stolen by spies
from the DNC's office at Whitehall. Their names betray their origins.

1 *Superbovitch* (Flagship)—1880—Box-battery ship—13 knots. Protection:
Battery, engines and magazines—C. Belt—D and F. Guns: 16 10 in.

2 *Captainski*—1870—Fully rigged turret ship—14 knots. Protection:
Turrets—D. Engines and magazines—E. Belt—F and G. Guns: four 12 in
in turrets, two 7 in.

3 *Temerairovitch*—1877—Box-battery and barbette ship—14 knots.
Protection: Barbettes and battery—D. Engines and magazine—E. Belt—F.
Guns: four 10 in and two 11 in in the battery, two 11 in on single
disappearing mounts in the barbettes.

4 *Royal Alfredski*—1861—Belt and battery ship—12½ knots. Protection:
Battery, engines and magazine—F. Belt—G. Guns: ten 9 in in broadside,
four 7 in chasers.

5 *Rupertoff*—1874—Steam ram—12 knots. Protection: Turret—B. Engines

The Russian Fleet steaming into action. The Flagship, Superbovitch, *is on the extreme
right.*

and magazine—C. Belt—D. Guns: two 10 in in one turret.

6   *Belle-Isleski*—1878—Steam ram—12½ knots. Protection: Battery—D. Engines and magazine—C. Belt—F. Guns: four 12 in guns firing from corner ports in the upper deck battery.

To Sir Trafalgar's mind the main strength of his Squadron lay in his big guns, the *Inflexible*'s 16-inchers and the 12-inchers of the *Devastation.* These guns were slow-firing and so would be at a disadvantage if the two forces mixed it in a close-range mêlée; so he decided to fight, as far as possible, a long-range battle with his Squadron in line ahead. His two turret ships were better able to fire ahead and astern so he placed *Devastation* in the lead and the *Inflexible* at the rear, with *Alexandra* and *Hercules* in the centre.

Prince Hanliovitch liked not the prospect of entering into a fire-fight with a force containing a 16 in gunner so he decided to force a ramming battle as quickly as possible. Such a plan was probably correct when the composition of the Prince's Squadron is considered, and accorded perfectly with his notoriously head-long approach to tactical matters.

He decided to attack in two arrowhead lines. In the first line were *Captainski* to port, *Temerairovitch,* whose barbette guns gave reasonable ahead fire, at the apex of the arrowhead in the centre, and his Flagship *Superbovitch* to starboard. In the second arrowhead were *Belle-Isleski, Royal Alfredski,* and *Rupertoff.* The *Royal Alfredski* had neither the ramming power nor axial fire really to warrant her position in the apex of this line but Hanliovitch regarded her as being a stately and handsome ship and thought that any position other than the van would demean her. Such aestheticism is unfortunately rare amongst warriors.

The battle opened with the British Squadron steaming in a northerly direction in line ahead and the Russians in their two lines to the north-east steaming west.

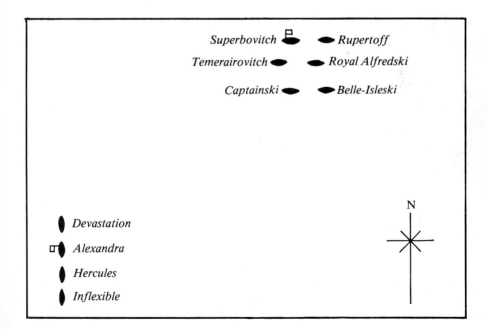

Sir Trafalgar saw that with the enemy so intent on coming to close quarters there was little chance of his long-range, stand-off fight so he prudently held his fire until he could be more sure of hitting. All guns of 9 in calibre and larger had been loaded with shot whilst the smaller guns, whose penetration was negligible anyway, were loaded with shell.

For a couple of minutes, whilst the air grew pregnant with tension and black with funnel smoke, the two forces steamed silently on towards each other. At last Sir Trafalgar judged the moment right and fired *Alexandra*'s whole starboard broadside at the *Captainski,* the nearest enemy, at which signal the rest of the squadron opened fire with all the guns which would bear, *Devastation*'s four turret guns, *Hercules'* starboard battery and bow guns and *Inflexible*'s two monster guns in the after turret (the offset forward turret being masked by superstructure). Unfortunately the range was still long and the gunnery poor so the *Captainski* suffered nothing worse than a good dousing as the shot and shell plunged into the water around her.

A minute or so later the two fleets were rapidly approaching close range and the Russians opened fire. *Superbovitch* fired her two bow 10 in guns at *Alexandra* at 250 yards range but missed, the *Temerairovitch*'s port battery and forward barbette 11-inchers also missed *Devastation*. The *Captainski,* however, had more luck, scoring three 12 in hits out of four on *Hercules*. All penetrated her rather thin armour and she suffered damage on the main deck ahead of the battery, the lower deck amidships and also suffered a penetration at the water line.

All the British big guns which could bear were being loaded, so that the only reply was a 7 in shell from *Hercules'* bows which burst against *Captainski*'s turret armour.

The British ships were in imminent danger of being rammed and it was none too soon that Sir Trafalgar hoisted his signal to turn individually towards the enemy. As they turned to starboard and steamed into the gaps between the leading Russian Ironclads the Russians had their chance to ram but they funked it, preferring to run past the British, firing off their broadsides as they bore. *Superbovitch* fired her portside 10 in guns at point-blank range into *Devastation* as she passed; only one shot hit and that penetrated the lower deck. On her starboard side the still unloaded *Devastation* was also subjected to passing fire from the *Temerairovitch* who fired two 10 in and one 11 in gun at her. One 10 in shot struck home and penetrated her boiler room flooding one of the furnaces and reducing her to three quarters speed. On *Temerairovitch*'s starboard side was the *Alexandra* who was fired at with the two loaded 10-inchers—both missed, however, despite the short range.

The *Captainski*'s guns had fired only a minute before and were still loading so she was unable to reply to the hail of fire to which she was subjected. A starboard 10-incher from *Alexandra* hit her steering gear and jammed the rudder in the ahead position. One 10-incher from *Hercules* wrecked her after turret, another smashed into her lower deck and a 7 in shell from a stern gun brought her funnel down. Immediately after a 16 in shot from *Inflexible*'s hitherto unengaged forward turret ripped into her forward magazine—a momentary delay and the *Captainski* erupted in a column of flame and splintered boats. Too busy to cheer, the port gunners aboard the Flagship laid their pieces on the *Temerairovitch* and were rewarded with a hit on the upper deck aft which failed to damage the after 11 in gun (as it was down below the

Above *The Russians open fire at short range.*

Below *As the two fleets comb each other* Captainski *is struck in the magazine and blows up.*

barbette loading), and a hit forward in the battery which destroyed the two 11 in guns there.

After steaming through each other the British Squadron and Russian first line both turned to re-engage, the British turning to starboard and the Russians to port. The Russian second line steaming on at 12 knots now attacked the turning British. *Royal Alfredski* poured her broadsides at point-blank range into both *Alexandra* and *Devastation* as she steered between them but she caused little real damage to *Alexandra,* and *Devastation*'s armour was proof against her old 9 and 7 in guns.

Belle-Isleski made a ramming run against *Hercules,* causing her to swerve dangerously across *Alexandra*'s bow, but she missed and suffered a 7 in and three 10 in hits from *Alexandra* and *Hercules* for her trouble, one of the latter smashing into her engine room and reduced her to two thirds speed. She fired three of her corner-mounted 12-inchers as targets appeared on her quarters, but all missed.

Meanwhile the *Rupertoff* put her helm over and ran down on Sir Trafalgar's apparently vulnerable Flagship as she turned. By so doing she put herself broadside on to the *Devastation* who, despite her boiler damage, was still travelling at 11 knots. She rammed *Rupertoff* square amidships and when the two ships separated the extent of the damage was revealed—22 sections of the *Rupertoff*'s hull were wrecked and flooding. It was enough to sink her and as her crew lowered her boats into the water she began to capsize. The *Devastation*'s ram was unsupported by armour and the impact had stoved her bow in by four sections; she could now manage no more than 6 knots.

Superbovitch's forward 10-inchers fired on *Inflexible* at 900 yards range, but

The fleets turn to re-engage as the Devastation *rams* Rupertoff *amidships.*

Royal Alfredski in the distance turns about and heads for the stationary Devastation *at full speed.*

the shot bounced off her impregnable citadel armour.

Devastation backed water at 2 knots to clear herself of the sinking *Rupertoff*. Her guns were now loaded, having been delayed in firing by the shock of ramming, but now they were laid upon the *Royal Alfredski* who, 300 yards away, was turning round towards her. They fired and two 12 in shot struck home, one devastating *Royal Alfredski*'s bows and chase guns and the other bursting into the main deck battery putting six of her 9 in guns out of action. *Royal Alfredski* replied with her bow guns and scored a hit on *Devastation*'s after funnel with a 7 in shell.

Alexandra, Hercules and *Inflexible* continued to steam south-east whilst *Temerairovitch* and *Superbovitch* raced towards them on a converging course and *Belle-Isleski* steamed along on their other side. *Hercules'* broadside was masked and *Inflexible*'s guns still loading, so only *Alexandra* was able to fire but all four of her guns missed the *Superbovitch* despite the range being only 150 yards. The Russian Flagship could not reply but *Temerairovitch* fired her forward barbette 11 in gun at *Alexandra* and hit her on the waterline aft. *Belle-Isleski*'s after starboard gun, the only one loaded, fired into *Hercules* and struck her in the lower deck below the battery.

The *Royal Alfredski* charged down upon the barely moving *Devastation* at 12 knots, intent upon ripping her open with her ram, but *Devastation*'s 2 knots gave her just enough way on to turn the Russian's bow. The two ships bumped and scraped sides and as they separated the Russian fired her two remaining port 9-inchers. Both shots hit, one broke up against the forward turret armour but the other broke into the breastwork amidships.

Temerairovitch also steamed full tilt at *Alexandra* but again the ram was foiled with a deft use of the helm. *Superbovitch* ran past the Flagship firing into her with her starboard broadside guns, hitting her with 10 in shot twice on the waterline, once on the lower deck and once in the poop. *Temerairovitch*'s port battery and after barbette guns penetrated *Alexandra* once more on the waterline and in the engine room. She was in a bad way and began to sink by a deck; fortunately her lower deck still gave her enough buoyancy to keep afloat, but her

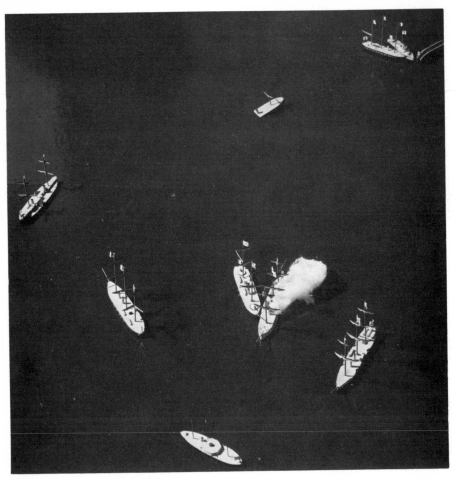

The Alexandra *is badly damaged and sinks by a deck.*

speed was much impaired. She replied by hitting *Superbovitch* twice on the main deck and *Temerairovitch* on the waterline and the main deck. *Hercules* also fired into the *Temerairovitch,* hitting her with 10 in shot in the waterline and lower deck and a 7 in shell on the upper deck.

Meanwhile, *Inflexible* was positioning herself for the moment when her guns would be able to fire, and the optimistic Captain of the *Belle-Isleski,* although reduced to 8 knots, followed in the hope of ramming.

Royal Alfredski, much reduced in firepower, steamed away from *Devastation* at full speed while the latter's guns were still in the loading position. *Devastation* was only chugging along at 4 knots and no firing took place between them.

Inflexible and *Hercules, Superbovitch, Temerairovitch* and *Belle-Isleski* continued on their circular, dog fight courses but *Alexandra* was now low in the water and reduced to 6 knots. Of the three Russian ships only the Flagship could maintain full speed, both the others suffering from engine room damage. The *Hercules'* after 9 and 7 in guns fired at *Temerairovitch* but the only one to hit, a 7 in shell, burst against her armour. *Royal Alfredski*'s two starboard 9-inchers

The Belle-Isleski *is reduced to a wreck by* Inflexible*'s heavy guns.*

fired upon *Inflexible* but missed. *Inflexible*'s after turret guns were now loaded and were aimed and fired at *Belle-Isleski* who was only 400 yards away on the starboard quarter. Both 16 in shot hit, the first striking below the waterline, flooding the engine and boiler rooms and the second smashing into the lower deck. The *Belle-Isleski* was reduced to a shambling wreck, down by a deck, dead in the water and almost sinking.

Prince Hanliovitch reviewed his position and found himself with one ship in good fighting order (his Flagship), one still steaming well with little fighting power *(Royal Alfredski)* one reduced in fighting power and steaming ability *(Temerairovitch)* and one in danger of sinking *(Belle-Isleski)*. He decided that the day was lost and that his mission had become impossible so he gave the order to retreat. Even before the signal was hoisted he showed leadership by example and had *Superbovitch* turn for the straits, and flee. *Royal Alfredski* and *Temerairovitch* followed as best they could; they were fired on by the *Devastation* and *Hercules* but Hague was not in the mood for a chase, being too pre-occupied with the parlous state of his Flagship. The *Alexandra*'s upper deck

The Russians flee.

guns fired on *Belle-Isleski* hoping to finish her off but the only hit smashed into the upper deck and left her still afloat. Her Captain eventually saved the British the trouble by abandoning the ship and opening her sea cocks.

With the Russians in retreat Sir Trafalgar Hague and his staff climbed into the *Alexandra*'s only remaining undamaged boat and were rowed across to the *Inflexible*. *Alexandra* and *Devastation* would have to be sent to Malta for extensive repairs but the *Inflexible* was completely undamaged and the *Hercules'* repairs could be effected by her own engineers without recourse to HM dockyards. He could continue his mission to Constantinople and, furthermore, negotiate with the Turks with the added prestige of having soundly thrashed the Ruskies. He would write a report to the Admiralty extolling the virtues of the fast-firing, hard-hitting 10 in muzzle-loader and the great merit of Cheltenham Spa water in rejuvenating tactical genius. After that he would give some thought to the Birthday Honours List in which he was sure to be mentioned. Viscount Hague of the Hellespont, he mused, would sound well in Cheltenham society.

Chapter 10

The Dreadnought period

This chapter is devoted to a set of rules for use in wargaming fleet actions in the Dreadnought era, which for all practical purposes can be taken to mean the First World War. The Dreadnought battleship, though, was still the prime unit of naval warfare up until the early 1930s and in many men's minds it continued to be so until Matapan and Pearl Harbor disabused them of the notion. These rules, then, can quite happily be used for battles in mythical wars right into the 1920s and '30s.

The rules were written with the specific object of allowing a single player per side to handle comparatively large fleets of model battleships. To make them manageable for large numbers of ships we have had to simplify them in certain respects and most of this pruning of detail has been applied to the light ships, the cruisers and destroyers, which although attached to all 'big ship' units in this era played only an auxiliary role in a fleet action. Should you wish to fight smaller battles and give more character to the small ships, however, or if you want to fight cruiser actions, then the rules can be adapted quite easily by applying to the light ships the same system of damage, etc, as used on the capital ships.

In addition to simplified rules relating to light ships, I also have them on a different scale of values for the purposes of floatation and damage calculations, which has the effect of preventing battleships and battlecruisers from firing on destroyers and cruisers and vice versa. Visiting wargamers often look askance at this feature of my rules but it serves a useful purpose, as well as being not too inaccurate (the guns of small ships would not penetrate the armour of capital ships, whilst big AP shells would often pass through unarmoured hulls without detonation). When calculating the smashing effect of shells upon the floatation value of a ship (based respectively on shell weight and displacement) it is easier, and far, far quicker to cross out three boxes (11 in shell) from a Dreadnought's 52 boxes (ie, 26,000 tons) than to do the mental sum of 26,000 minus 660 lb. When light ships are included the possible range of shell weights (31 lb for a 4 in to 3,200 for an 18 in) and displacements (600-ton destroyer to a 35,000-ton Dreadnought) are so great that the system does not work with its intended simplicity, so I use two scales of value.

The ships

The armoured battleship reached the peak of its power in the very early years of this century when the firing of big guns had become a science and when mines and torpedoes were still immature novelties. By the First World War the battle-ship was still queen of the seas and still the unit by which a nation's naval might was measured, but the danger from torpedoes and mines was becoming more real and the battlefleet's freedom of movement was consequently reduced by the dictates of caution. The battlefleet had ceased to be simply a force of capital ships fearing no opposition except from other capital ships; it had become an 'all arms' division of ships including strong cruiser and destroyer elements whose duties were to reconnoitre for the battleships and to protect them from torpedo attack.

The hardcore of the battlefleet, around which the whole force was built, was, of course, the Dreadnought battleship. The Dreadnoughts were of between 20,000 and 30,000 tons displacement, usually turbine-propelled and capable of 21 knots or more, but the feature which characterised them and distinguished them from their pre-1906 'pre-Dreadnought' predecessors was their armament. Unlike the earlier ships which had a main armament of two or even three calibres, say four 12 in, four 9.2 in and eight 6 in, the Dreadnoughts had a main battery of all big guns, eg, ten 12 in; any 4 or 6 in guns carried were purely for defence against torpedo boats. The purpose of the one-calibre battery was not, strangely, to hit with more big shells but to allow the hitting to be done at longer ranges. When a mixed-battery ship of the old type fired, the large number of splashes and the different flight times of the various sizes of shells prevented the gunners easily recognising their own shellburst and so hampered correction. In the Dreadnoughts a single officer in the control or director position up on a mast controlled, fired and spotted for the whole main battery, which was fired as two salvoes, so that spotting was made easy and ranges of up to 20,000 yards became possible. The object of this officer was to correct his salvoes until he had a tight group of shell splashes straddling the target, some under, some over. When this happened rapid salvo fire commenced, with a proportion of the shells hitting the target, until the range was lost, when spotting and correction began again.

The battlecruisers were a controversial breed, but before their weaknesses were highlighted in the hard school of war they were generally regarded, at least by the public, as the epitome of dash and *élan,* as the cavalry of the seas. They were similar to the Dreadnoughts but of greater speed, bought at the expense of armour and gun power. In the British service the main guns were usually of the came calibre as the contemporary battleships but with one turret less; in the German service they were usually smaller (11 in against 12 in) but the same number. Their duties, when attached to the battlefleet, included reconnaissance, their enormous firepower allowing them to fight their way through the enemy's cruiser screens in search of his battlefleet, and acting as a fast van to the Dreadnoughts which would attack, turn the head of the enemy line and prevent his retreat. To this last role their thin armour made them particularly unsuited and in the Royal Navy it was taken over by fast battleships when these became available.

Although the introduction of the Dreadnought type rendered the old mixed-battery battleships obsolete there were so many of them in existence that they continued to serve right through until the end of the war. The Royal Navy had 40 of them in 1914 and the Germans 23 and they were used by both şides in the early years of the First World War to pad out the main battlefleets until sufficient Dreadnoughts became available. Their light broadsides, however, made them unsuitable for use against enemy Dreadnought fleets and their low speed was a handicap to their own fleet, so that they were eventually relegated to side-show waters, such as the Mediterranean. As the Dreadnoughts rarely strayed from the North Sea, though, they were still a force to be reckoned with in these secondary areas of naval activity.

In the days of the mixed-battery battleship armoured cruisers had filled the same role as did the battlecruisers in the Dreadnought fleet, but their armament was usually of a smaller calibre than their battleship comtemporaries'. They were relatively slow, of little fighting value, and expensive to operate but like the old battleships they continued in service because there were so many of them. Eight of them went to Jutland with the Grand Fleet but only five came back.

Light cruisers were small ships, usually unarmoured and carrying guns of 4 or 6 in calibre. In the wider scope of imperial strategy they were used to patrol the sea lanes, protecting friendly commerce and attacking that of the enemy. When with the battlefleet they steamed ahead and on the flanks reconnoitring and screening the fleet from the enemy's cruisers. Fighting battleships was no part of their duty and my rules do not really allow them to.

Destroyers during the First World War were still quite small vessels being usually of less than 1,000 tons displacements. The Germans still designated their destroyers as torpedo boats and they were indeed built around the torpedo as the prime weapon. The British were more committed to the big gun battleship and so their destroyers were designed as protectors of the battlefleet from enemy torpedo boats. They were consequently built bigger than the German boats, with greater firepower but a smaller torpedo armament. Destroyers and torpedo boats operated in loose formation within the flotilla and covered the battlefleet during deployment or retirement.

Tactics

The vast number of ships comprising the Grand Fleet and the long distances at which the ships fought led to the belief that a battle fought by sub-units of battle squadrons or divisions would be impossible for the Commander in Chief to control and co-ordinate. As a result the battleships operated as a single indivisible body and reverted tactically to the single line ahead commanded from the Flagship in the centre. As a result doubling and other methods of dual attack could not be used.

The actual physical concentration of ships was no longer strictly necessary as gun ranges had become so long that concentration could be achieved by gunfire alone. Such fire concentration, however, was rarely practised as it was found that spotting became confused and consequently accuracy reduced, when a number of ships fired at the same target. At the same time those enemy ships which had been left unengaged became dangerously accurate as their gunnery control teams made cool and calm corrections undisturbed by shell hits and near misses. The normal firing system, therefore, was for each ship to engage its opposite number.

Single lines ahead with ships engaging their opposite numbers had led to indecisive combat in the 18th century and the disappointing results of Jutland, the only fleet action of the war, suggest that the system had not improved with age. A practical method of concentration, and one successfully employed by Jellicoe at Jutland, was bringing the battle line across the head of his line at right angles to its course so that they could not reply, ie, crossing the T. This tactic had been commonly used by individual broadside sailing ships but the longer ranges of 20th century guns and the wider arcs of bearing of turret guns made it a practicable proposition for use by one fleet against another.

In order to allow the battlefleet to form line of battle on any course in the minimum time the fleet cruised with the battleships in columns of divisions with columns disposed abeam. On being informed by his advanced cruisers of the course and position of the enemy fleet the CinC would turn the divisions in succession to port or starboard as appropriate and thus bring them into a line ahead. Any acute turns by the line were best executed before the enemy came within range because when turning in succession the enemy would quickly find

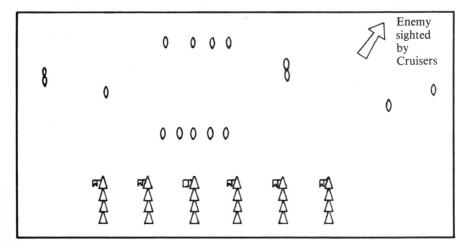

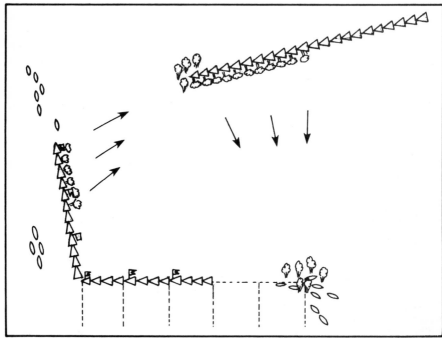

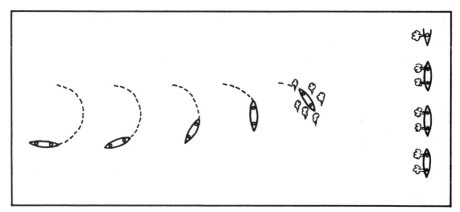

Left *The British Battlefleet in columns of divisions in line abreast preceded by its cruiser and armoured cruiser screen. Prior to deployment at Jutland.*

Below left *Jellicoe has ordered a deployment on the port column, thus bringing his line across the T of the German line. His deployment has been delayed, though, and his line is turning in succession within range of the enemy guns. All the rear ships will be subjected to some uncomfortably accurate salvos as they pass through 'windy corner'.*

Above *The* Gefechtskertwendung, *or battle turn away. Turning in succession in these circumstances would have been disastrous.*

the range of the corner and give the rearward ships of the line a severe mauling as they turned. Jellicoe left his otherwise masterly deployment at Jutland a little late and his rear ships went through a very uncomfortable few minutes as they passed through 'windy corner'.

The Germans were always aware of the enormous British superiority in numbers and knew that if their battlefleet met the British they might be compelled to withdraw under fire. If this were done in succession the result would have been catastrophic, so they developed and regularly practised a very difficult manoeuvre by which a line could reverse direction quickly and without turning in succession. The *Gefechtskertwendung,* or in English, 'battle turn away', involved the last ship in the line putting its helm over and turning through 16 points (180 degrees). As the ship started to turn the next ahead followed suit until the whole line was steaming the way it had come with the rear ship leading and the leader at the rear. Scheer used the manoeuvre successfully twice at Jutland but on the third occasion when the line was slightly bent and the ships' shell battered it, there was something of a shambles with engines being thrown astern to avoid collision.

The Rules

Model and sea scale

The most popular scale of models of this period, 1:1,200, is far too large for use in wargames where ranges of up to 20,000 yards have to be represented. Models of the smallest possible scale should be used, and this is 1:4,800. My forces, however, are in 1:3,000 scale as the available range of models is much bigger, and so it is around this scale that these rules are written.

The scale 1:3,000 sounds minute, but even so 20,000 yards converts into an impossibly long 20 ft. This scale must be 'adjusted' (or fiddled) so that the battle can take place on the floor of a normal-size room but without altering the ratio between range, length of line of battle and speed. If we let these ratios change then the number of ships within range and able to engage will be incorrect and the volume of fire which an approaching or retiring ship can be subjected to will be wrong, so that the advantages of superior numbers and better tactics are reduced or eliminated.

My solution to this problem is as follows:

Firstly: the problem is halved (literally) by using one model ship to represent two actual ones of the same type. This halving is facilitated by the fact that both the British and Germans built their capital ships in classes of four and organised them into divisions of four and squadrons of eight. Thus a model fleet will contain, as far as possible, one model for each two real ships of that class and will be organised into divisions of two and squadrons of four; two fleets organised in this manner will therefore be of exactly the same relative strength as their real-life prototypes.

Secondly: the model capital ships are mounted on card bases 3 in long. The scaling is calculated not from the models but from the card, which represents the amount of sea room in the line which the *two* battleships would occupy. Dreadnoughts usually required 750 yards each in the line, so the two ships represented by our model would occupy 1,500 yards, which is represented by our card base.

The scale, then, is 3 in = 1,500 yards or 1:18,000 giving us a maximum range of 40 in for a 20,000-yard gun and 7 in movement per five minute period for a 21-knot ship. Much better!

Above right *A battle squadron or two divisions. On the bottom row are the eight real Dreadnoughts and on the top row the four models representing them.*

Below *A model mounted on its card base, which helps identify the ship by bearing its name and speed.*

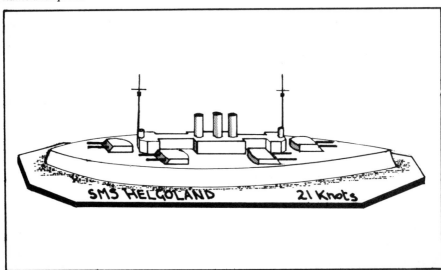

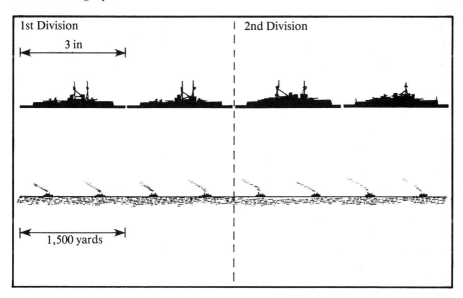

Sequence of play

1 Orders (ie, signals) to units of the fleet are written down, to become executive next move.
2 Ships are moved in accordance with the last order issued. Torpedoes may be launched at any point in the move, to run a proportion of this move equal to the amount of 'move' left after its launching.
3 Guns are fired. All firing is simultaneous so that any damage caused does not take effect until next move.

Movement

Each movement period represents five minutes of real time. At our sea scale of 1:18,000 a ship will move $\frac{1}{3}$ in for each knot of speed.

Speeds may be increased or decreased by a maximum of 5 knots per move for capital ships or 8 knots per move for cruisers and destroyers. Speed reductions caused by damage will take effect immediately.

Armoured cruisers, battleships and battlecruisers will turn on a $1\frac{1}{2}$ in circle; destroyers and cruisers on a 1 in circle.

If the maximum speeds of actual ships are not known the following can be taken as being typical for each type:

Dreadnoughts	—	21 knots
Battlecruisers	—	27 knots
Pre-Dreadnoughts	—	18 knots
Armoured cruisers	—	23 knots
Light cruisers	—	28 knots
Destroyers	—	32 knots

Formations

The smallest sub-unit of battleships is the division which in reality was usually four ships, and in our reduced scale is two. The leading ship of the division is assumed to be the Rear-Admiral's Flagship and this ship must always lead the other, in a two-ship line ahead. The only occasions this rule is relaxed are (a) when the Germans (not British) carry out a *Gefechtskertwendung* and (b) when the ships turn out of line to comb torpedo tracks. In this latter case they must resume divisional line, Flagship leading, as soon as possible.

Battlecruisers, fast battleships and armoured cruisers did not always operate in such large numbers as to make a breakdown into equal-size divisions, but they should always conform, as far as possible, to the line ahead rule with divisional or squadron Flagship leading.

Because one ship represents two real ones and two ships a division the advantage the Germans should have in being able to perform the *Gefechtskertwendung* over the British who are restricted to a 180 degree turn by division is diminished. To make the British more clumsy in this respect and thus restore the balance, we insist upon the British divisions turning about in two moves, each of a 90-degree turn, so on the first move the division will turn by eight points (90 degrees) and travel to the end of their move, and then on the second move they must travel at least half their move before turning the other eight points.

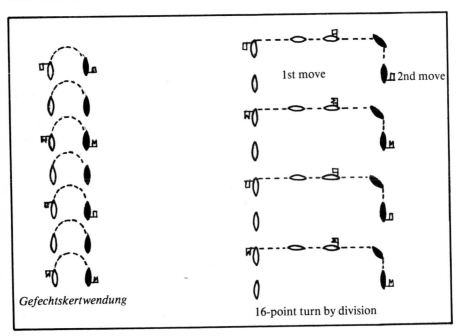

Gefechtskertwendung

16-point turn by division

Light cruisers and destroyers may adopt much looser and less rigid formation than capital ships and are not subject to the Flagship leading, or '180 degrees in two moves' rule. Light ships may also pass through the gaps between capital ships in the line, because of course the real spaces are much larger then we have represented with our models.

Firing

In the five-minute period represented by our game move a battleship's big guns would have been able to fire about ten salvoes, and the smaller, quick-firers even more. It would take far too long if we attempted to calculate the effect of every salvo, so firing is carried out only once per move but the hitting probability and destruction value of shells reflect this and also the faster rates of fire of smaller guns.

A Dreadnought's big guns were too slow in training and rate of fire to be of much use against fast-moving light ships, and their shells were likely to pass through their sides without exploding. At the same time 4 and 6 in guns were unable to penetrate battleship armour, so to quicken up the game and to facilitate our simplified damage record system we do not allow guns bigger than 9.2 in to fire at light ships, or guns smaller than 7.5 in to fire on capital ships. Our damage record system means that the intermediate calibres which can engage both types of ship must have two scales of destruction value.

Destroyers and cruisers which are between the firing ship and its target may be fired over providing they are at least 4,000 yards (8 in) from both firer and target. Their smoke, however, which is reckoned to trail 3 in behind them, will fog up the sighting and reduce accuracy.

The ranges of guns are as follows:

11 in and greater	—	20,000 yards (40 in)
7.5 in to 10 in	—	16,000 yards (32 in)
Less than 7.5 in	—	12,000 yards (24 in)

Firing procedure

1 Measure the range from forefunnel to forefunnel.
2 Fire guns of different calibres separately.
3 Throw two average dice per calibre, per ship, and add and subtract factors from the table below.
4 Compare the dice score with the number of guns fired in the appropriate range column of the gunnery results chart for the number of hits scored.

Fire factor table

Subtract 1 — For each ship whose smoke obscures the range.
If the firing ship has changed course by 45 degrees that move.
If new target that move.
For each ship over three firing at the same target.
Subtract 2 — If director is destroyed.
If firing at night.
Add 1 — If firing ship *not* under fire.
If target has turned in succession that move.
Add 2 — If target has turned in succession that move but the enemy line was also turning last move.

Simple firing system

This firing system lacks the realistic details of that just explained but it is much quicker to use and can be useful where very large fleets are employed.

Throw one *ordinary* dice per *two* guns firing. The table below indicates whether a hit has been scored or not.

	11 in +	7.5 to 10 in	Less than 7.5 in
4,000 yards (8 in)	2	2	2
8,000 yards (16 in)	3	3	4
12,000 yards (24 in)	4	5	6
16,000 yards (32 in)	5	6	—
20,000 yards (40 in)	6	—	—

	4,000 yards Dice									8,000 yards Dice									12,0 Di			
Guns	**3**	**4**	**5**	**6**	**7**	**8**	**9**	**10**	**11**	**3**	**4**	**5**	**6**	**7**	**8**	**9**	**10**	**11**	**3**	**4**	**5**	**6**
1	–	–	1	1	1	2	2	2	2	–	–	–	1	1	1	2	2	2	–	–	–	–
2	–	1	1	2	2	3	3	3	4	–	–	1	1	2	2	3	3	4	–	–	–	1
3	1	1	1	2	2	3	3	4	4	–	1	1	2	2	3	3	3	4	–	–	1	1
4	1	1	2	3	3	4	5	6	6	–	1	1	2	2	3	3	4	4	–	–	1	1
5	1	1	2	3	3	4	5	6	6	–	1	1	2	2	3	4	4	4	–	1	1	1
6	1	1	2	3	3	4	5	6	7	–	1	1	2	2	3	4	4	5	–	1	1	2
7	1	1	2	3	4	4	5	6	7	–	1	2	2	3	3	4	4	5	–	1	1	2
8	1	1	2	3	4	5	5	6	7	1	1	2	2	3	3	4	5	5	–	1	1	2
9	1	2	2	3	4	5	6	7	7	1	1	2	2	3	4	4	5	6	–	1	1	2
10	1	2	2	3	4	5	6	7	8	1	1	2	3	3	4	5	6	6	1	1	1	2

Gunnery results chart

Damage

Having fired our guns and, we hope, hit the target, we have to calculate the damage and its effect on our opponent's ship. Again this can be done in two ways, a quick and easy way, the 'progressive damage' method, or the more interesting but also more time-consuming 'critical hit' method: I prefer the latter system but your choice will depend upon taste and the size of fleet involved.

Progressive damage method

Each shell will destroy a number of floatation 'boxes' according to its size.

Standard damage

Shell	On capital ships	On light ships	Shell	On capital ships	On light ships
			10 in	2	—
18 in	15	—	9.2 in	2	6
15 in	9	—	8.2 in	1	5
13.5/14 in	6	—	7.5 in	1	4
12 in	4	—	6 in	—	2
11 in	3	—	4 in	—	1

As a ship's floatation boxes are destroyed it will lose speed and firepower accordingly.

Boxes destroyed	Guns able to fire	Speed
$\frac{1}{2}$		
$\frac{1}{2}$		$\frac{3}{4}$
$\frac{3}{4}$		$\frac{1}{2}$
$\frac{7}{8}$	Nil	$\frac{1}{4}$
All	Sinks	

Critical hit system

For every hit on a capital ship by a big shell, 11 in and larger, a dice is thrown. A throw of 5 or 6 indicates a critical hit, 1, 2, 3 or 4 a standard hit, in which case floatation boxes are crossed out in the same way as in the earlier method. Secondary guns and speed, and all light ship factors, are reduced by a progressive method which is explained in the next section.

For every critical hit throw a further dice to determine the part of the ship hit.

rds			16,000 yards Dice									20,000 yards Dice									Guns
9	10	11	3	4	5	6	7	8	9	10	11	3	4	5	6	7	8	9	10	11	Guns
1	1	2	-	-	-	-	-	-	-	1	1	-	-	-	-	-	-	-	-	-	1
2	2	2	-	-	-	-	-	-	1	1	1	-	-	-	-	-	-	-	-	1	2
2	2	2	-	-	-	-	-	1	1	1	1	-	-	-	-	-	-	-	1	1	3
2	2	2	-	-	-	-	1	1	1	1	1	-	-	-	-	-	-	1	1	1	4
2	3	3	-	-	-	1	1	1	1	1	1	-	-	-	-	-	-	1	1	1	5
3	4	4	-	-	-	1	1	1	1	1	1	-	-	-	-	-	1	1	1	1	6
3	4	4	-	-	-	1	1	1	1	1	2	-	-	-	-	1	1	1	1	1	7
3	4	4	-	-	1	1	1	1	1	2	2	-	-	-	1	1	1	1	1	1	8
4	4	4	-	-	1	1	1	1	2	2	2	-	-	-	1	1	1	1	1	1	9
4	4	5	-	1	1	1	1	1	2	2	2	-	-	1	1	1	1	1	1	1	10

Critical hits do not cause standard damage and for simplicity no account is taken of the greater smashing effect of bigger shells.

6 } 5 }	—	Turret
4	—	Bridge and conning tower destroyed
3	—	Director/gunnery control position destroyed
2	—	Rudder hit
1	—	Magazine hit

Turret—Throw a dice to determine which turret has been hit—its guns are put out of action for the remainder of the battle. On pre-Dreadnoughts with only two turrets, give them an equal chance of hitting the foreturret, the after turret and of scoring a standard hit instead.

Bridge/Conning tower—The ship may not change course for three moves. The ship will not respond to any orders during that period and if it is a Flagship it will not issue any.

Gunnery control hit—Aiming and spotting is carried out from the individual turrets. Deduct 2 from the firing dice.

Rudder—Throw a dice. 1, 2 rudder jammed to port and ship circles on a 3 in circle until the rudder is freed. 3, 4 rudder jams in the straight ahead position and the ship may not change course. 5, 6, rudder jams to starboard and the ship circles. Throw a dice every subsequent move to attempt to free the rudder, 5, 6 mean success.

Magazine—The shells may explode the magazine and destroy the ship, or it may just fire the charges in the turret shaft. In this latter case the ship will suffer ten points standard damage and will lose a turret (dice for which). A dice is thrown to determine the effect of magazine hits:

Ship type	Magazine explodes ship destroyed	Turret destroyed 10 points damage
British armoured or battlecruiser	1, 2, 3, 4	5, 6
British battleship	1, 2	3, 4, 5, 6
German ship	1	2, 3, 4, 5, 6

Damage record cards

The damage record card explained here presumes the use of the critical hit system of gunnery; it will soon become apparent, though, what alterations are necessary for use with the simpler system.

As I have mentioned earlier we use a different scale of values for light ships and we do not apply the critical hit system to them, so that their damage record cards are different to those for armoured ships. The system here described relates to armoured cruisers and bigger ships, destroyers and light cruisers being dealt with later.

Award to each ship two floatation boxes per 1,000 tons displacement and two boxes per inch of armour at its thickest point on the belt. German ships were built with considerable emphasis on internal sub-divisions and damage control so we increase their floatation boxes by 20 per cent.

Turret guns are reduced by critical hits but secondary guns on the maindeck are reduced proportionately with standard hits. Divide the number of such guns which can bear on a broadside into the total number of floatation boxes, so that as that number of boxes is destroyed the number of secondary guns is reduced by one.

Speed remains unaffected until half the boxes are destroyed when speed falls pro rata so that the ship is dead in the water just before it sinks. Divide half the number of boxes by the ship's maximum speed in knots and when half-damage has been suffered reduce the ship's speed in units of 3 or 4 knots when the appropriate number of floatation boxes have been destroyed.

A rough plan of the ship will be needed so that turrets, etc, can be crossed out as they are destroyed by critical hits.

Here, then, is an example of a damage record card for SMS *Westfalen,* a German Dreadnought of the first class laid down. She displaces 19,000 tons, has an 11½ in belt, mounts 12 11 in guns and 12 6 in and has a maximum speed of 20 knots.

19,000 tons @ 2 boxes per thousand =	38
11½ in of armour @ 2 boxes per in =	23
	61
Plus 20 per cent for superior German damage control	13
Total floatation boxes	74

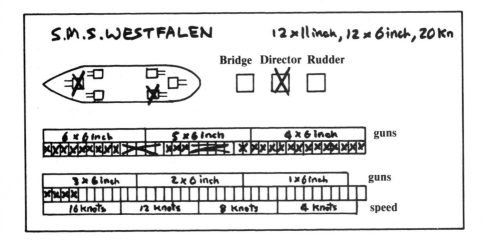

There are six 6 in guns per broadside, so one will be lost per 12 floatation boxes destroyed. Speed will reduce by 4 knots for every nine boxes destroyed over a half.

We can see on the chart that *Westfalen* has lost two turrets and her director position from critical hits. She has also suffered 41 standard damage points and in consequence is down to three 6 in guns per broadside and 16 knots speed.

The damage record cards for light ships are similar to those for capital ships except that we award *ten boxes* per thousand tons displacement, we do not give a 20 per cent bonus to German light ships and we do not use a plan view of the ship (because we do not use the critical hit system). It is necessary though, to become familiar with the layout of the ships in order to determine the number of guns able to fire on each bearing. Torpedo tubes are reduced proportionately, like guns.

Here are two examples of light ship damage record cards, one for a cruiser and the other for a destroyer.

1   HMS *Bellona*. British scout cruiser, 3,400 tons, 26 knots, ten 4 in guns (five per broadside). Thus, 34 floatation boxes, lose one gun for each seven destroyed and lose 3 knots per two boxes destroyed over 17.

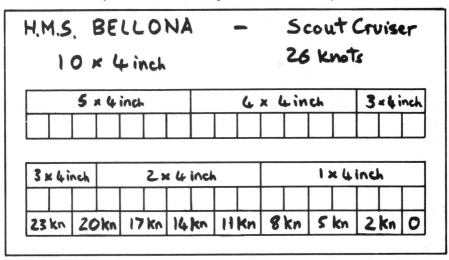

2   *G85* German destroyer, 1,000 tons displacement, 34 knots, three 4 in guns, six torpedo tubes. Therefore, ten floatation boxes with one gun and two torpedo tubes lost per three boxes destroyed and a 7-knot speed reduction for each box lost after five.

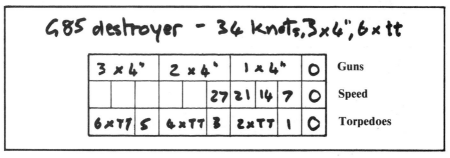

Night fighting

Before the Battle of Jutland the British were not trained in, or equipped for, night fighting and suffered for their neglect. Night fighting is very difficult to simulate in a wargame, particularly if no umpire is present, but it does sometimes occur in a campaign, so a set of rules are included here.

1 Orders are written out for *four* moves ahead and no units may react to things occurring outside their range of visibility.
2 Visibility is 2,000 yards or 6,000 yards if the target is illuminated by searchlight or star shell. Destroyers were usually painted black at this period so that if they are travelling at less than 10 knots (to supress funnel sparks) they are only visible at 1,000 yards, or 4,000 yards illuminated.
3 Guns may not be fired beyond the range of visibility and when fired *within* the range of visibility 2 is deducted from each dice throw.
4 Ships using searchlights reveal themselves to all other ships within 6,000 yards, but they may not be fired upon by the ship held in the beam as the gunners are dazzled. The German searchlight control was greatly superior to that on British ships so that where two ships turn searchlights on each other we assume the British gunners and light operators to be dazzled and temporarily out of action.
5 Only the Germans may fire star shells. They may nominate any position up to 6,000 yards away where the shell will burn overhead for one move, illuminating everything within 500 yards radius.
6 In any interchange of fire at night the normal rule of simultaneous firing does not apply. The ship which switches on a searchlight or fires a star shell first will always fire first and inflict its damage before the enemy can reply. Where illumination is not used the Germans will fire first.

Torpedoes

Almost every warship active during the First World War was equipped with torpedo tubes but so few were fired by cruisers or capital ships that in our wargame we can reasonably limit their firing to destroyers.

Torpedoes may be launched at any point during a move and any number may be launched at one time up to the number of tubes carried (I usually standardise British and German destroyers at two and four tubes respectively).

The torpedoes are represented by a marker, say a piece of matchstick, which runs to a range of 8,000 yards (16 in) in a move and then sinks. If, then, the torpedo is launched half way through a move it will run 4,000 yards this move and 4,000 yards the next before sinking half way through the move. The marker and any ships which look like being in danger from it are moved in phases until hits or misses are confirmed.

That a marker comes into contact with a model ship does not necessarily mean that a torpedo has hit, as our ships are six times oversize (or, more properly, three times as there are two of them). For each torpedo of the batch which the marker represents throw a dice and count a 5 or 6 as a hit, and anything lower as a miss.

For every torpedo hit throw two average dice and multiply the total by four. The resultant figure is the number of floatation boxes destroyed. Ships which are fitted with anti-torpedo bulges only suffer half this damage.

Mines

Mines were used defensively when they were sown in fields around the coastline with clear channels left for use by friendly ships and offensively when they were dropped outside enemy ports, or across the anticipated route of enemy ships, by submarines and fast minelayers. These latter minefields would have been rather haphazardly laid and so were less destructive than the regularly laid defensive fields.

When a ship enters a defensive minefield it throws one dice per 5 knots speed per move, so that a Dreadnought blundering into a field at 21 knots will have four dice thrown for it, a 1 indicating a mine explosion. In offensive minefields one dice is thrown per 10 knots speed. Having struck a mine a destroyer or cruiser will sink. Capital ships will have two average dice thrown for them and the total, multiplied by eight, will be the number of floatation boxes destroyed.

Submarines

Obviously submerged submarines cannot be represented in our wargames on the floor, but they do have an important role in the naval strategy of the First World War so we need a set of rules whereby submarines can be employed in a map game or campaign.

The submarines of that era were very slow and hunted by lying in ambush, submerged, rather than by pursuing their prey. For this reason we assume submarines to be stationary on the map, lying submerged during the day and both unseen and unseeing at night. Their endurance is 14 days, so the length of time they spend on station is 14 days less the outward and return journey time at 8 knots. When a ship passes within 15,000 yards of a submarine an attack may be made. At the same time if the target ship is escorted by destroyers and the submarine reveals itself by attacking, the destroyers may counter-attack. Two average dice are thrown to determine the success of the submarine's attack, and then for the destroyer counter-attack if this is made. Add 2 to the submarine's attack dice if the target ship is moving slower than 5 knots, add 1 if it is moving slower than 10 knots and deduct 1 if it is moving faster than 15 knots.

Submarine attack table

Dice	5,000 yards distance	10,000 yards distance	15,000 yards distance
10	2 torpedo hits	1 hit	1 hit
9	1 hit	1 hit	1 hit
8	1 hit	1 hit	miss
7 or less	miss	miss	miss

Destroyer attack table

	5,000 yards distance	10,000 yards distance	15,000 yards distance
1–2 destroyers	8–10 sinks sub	9–10 sinks sub	10 sinks sub
3–5 destroyers	7–10 sinks sub	8–10 sinks sub	9–10 sinks sub
6+ destroyers	6–10 sinks sub	7–10 sinks sub	8–10 sinks sub

Fuel

Fuel consumption and range is unimportant in the one-off wargame, but it can be of crucial importance in a campaign. The bunker capacities given are average for each type, as is the consumption. Some British ships were purely oil-fired which gave them a longer range than the coal-burners but their operational radius was limited to that of their consorts in the fleet, so all are treated as though coal-burners. German armoured ships, being designed for operations in the North Sea and Baltic, had a smaller bunkerage than the wider-ranging British ships.

Bunker capacity (tons)

Type	British	German
Battleship	2,000	1,700
Battlecruiser	2,500	2,200
Armoured cruiser	1,500	1,400
Cruiser	800	800
Destroyer	300	300

Hourly consumption (tons)

	In harbour	10 knots	18 knots	21 knots	25 knots
Battleship	1	4	11	18	—
Battlecruiser	1	5	13	20	30
Armoured cruiser	$\frac{2}{3}$	3	8	15	20
Cruiser	$\frac{1}{2}$	$1\frac{1}{2}$	4	7	10
Destroyer	$\frac{1}{4}$	$\frac{1}{2}$	$1\frac{1}{2}$	2	3

It takes 12 hours fully to coal a ship in harbour, or a pro rata period for a lesser quantity. Coaling at sea from fleet colliers takes twice as long and coaling at sea from other vessels or captured merchantmen takes three times as long.

Chapter 11

The Battle of Texel, 1916

The demonstration battles described in earlier chapters were kept small and simple so as to make the narration of the events easier to follow. For the demonstration of the Dreadnought rules, however, we are going to stage something more ambitious. The Battle of Texel, was fought out by model Grand and High Seas Fleets, scaled down from the strengths they had in mid-1916. The capital ships are on a scale of one model per two ships (although we give the Germans one additional battlecruiser to level the odds slightly) and the cruisers and destroyers on a scale of one to three or four.

The German Navy was at such a disadvantage numerically during the First World War that if we had followed our earlier practice of starting the battle by laying the rival fleets down in the playing area in sight of each other the result would have been an inevitable annihilation of the High Seas Fleet by the bigger-gunned and more numerous British. To avoid this the wargame was organised as a small map campaign in which the German Fleet would be given the opportunity to attack British coastal towns and minor naval units whilst evading the main force.

To ensure secrecy of movement and to allow an element of 'fog of war' the game was organised and controlled by an umpire (me). The two players were each given a small map of the North Sea, but these were for reference purposes only; sailing instructions, in note form, for each of their forces were issued to the umpire and he plotted the fleet movements, hour by hour on his map, which was a large-scale one of 1 in to 20 miles. The umpire was also given diagrams showing the cruising formations of the various forces, which showed with particular detail the positions and distances of the screening cruisers. As forces came within sight of each other the umpire passed to the players (by note) such information as was visible from the sighting cruisers eg smoke and bearing at 25 miles, type of ship (ie, large and small) at 20 miles, and ship class, numbers and course at 15 miles.

The map moves at this stage were made for five-minute periods to allow for quicker reactions to sightings and at the end of each five minutes the investigating cruisers would report back with more information whilst the Flagships (ie, the players) issued suitable orders to the ships of the squadron or fleet. When the screening cruisers came within gun range of each other, or when the main forces came within sight, the umpire laid the models out on the floor in their correct relative positions and formations and the players would fight out the battle. The movements of the other formations would still be plotted on the map each hour (12 game moves with the models), and the fighting forces would be returned to the map as soon as the two sides lost sight of each other.

The 'fog of war' was brought into the campaign in three ways:

1 The fleets' positions as plotted by the players on their small reference maps rarely corresponded exactly with their positions on the definitive umpire's map, so that orders were sometimes given on a false apprehension of the situation.

2 The umpire had planted neutral merchant ships around the map, so that their smoke, when sighted, would cause forces to run off on wild goose chases until they were identified at 15 miles distance.

3 Unbeknown to the German player, the umpire had told the British player of the approximate area of the German raid and had allowed the British a four-hour start over the Germans. This was intended to represent the advantage which the Admiralty had throughout the First World War of being able to intercept and decipher German naval radio messages and thus get forewarning of operations.

The German player had been given the objective of bombarding any British town on the East Coast. No points system for determining success had been devised, but at the end of the day when awarding the laurels the umpire would take into consideration the length and weight of the bombardment, the military importance of the target city, losses of ships and changes in relative naval strength resulting from the action.

Scheer selected Dover as his target, and gave the following reasons for his choice.

1 It was very far south and so allowed a good chance of evading the Grand Fleet based up in the Orkneys.

2 It was an important naval base and a key port in communication with the armies in Flanders.

3 It was protected by a fleet of old pre-Dreadnoughts which would be 'easy meat' for his battlecruisers and Dreadnoughts. These old ships, if destroyed, would have to be replaced by withdrawing Dreadnoughts from Scapa Flow, thus changing the naval balance in the northern waters in favour of the Germans.

The German Fleet was based in the Jade Bay and consisted of the following forces. The number of big guns per broadside is shown by each capital ship; where this is different from the total number this is shown in brackets.

In the following lists DB = Dreadnought, BC = battlecruiser, FDB = fast battleship (25 knots), B = pre-Dreadnought, AC = armoured cruiser, LC = light cruiser, TB = torpedo boat/destroyer, D = destroyer.

The Scouting Force— Rear-Admiral Hipper

1st Scouting Group		2nd Scouting Group
Derflinger (Flag) BC	8 × 12 in guns	*Regensburg* LC
Seydlitz BC	10 × 11 in guns	*Stralsund* LC
Moltke BC	10 × 11 in guns	*Graudenz* LC
Von der Tann BC	8 × 11 in guns	**1st Flotilla**
		4 TB

The Battlefleet—Vice-Admiral Scheer

1st Battle Squadron

Helgoland	DB	(12)	8 × 12 in
Westfalen	DB	(12)	8 × 11 in
Nassau	DB	(12)	8 × 11 in
Posen	DB	(12)	8 × 11 in

2nd Battle Squadron

Pommern	B	4 × 11 in
Hessen	B	4 × 11 in
Rheinland	B	4 × 11 in

3rd Battle Squadron

Friedrich der Grosse	DB	10 × 12 in
Kaiser	DB	10 × 12 in
König	DB	10 × 12 in
Grosser Kurfürst	DB	10 × 12 in

3rd Cruiser Squadron

Emden	LC
Leipzig	LC
Dresden	LC

4th Cruiser Squadron

Nürnburg	LC
Breslau	LC
Magdeburg	LC

2nd 3rd and 4th Flotillas
Each 4 TBs

The British forces were distributed along the whole length of the East Coast.

The Grand Fleet—Admiral Sir J. Jellicoe

At Scapa Flow
1st Battle Squadron

Orion	DB		10 × 13.5 in
Temeraire	DB	(10)	8 × 12 in
Revenge	DB		8 × 15 in
Neptune	DB		10 × 12 in

4th Battle Squadron

Iron Duke (Flag)	DB		10 × 13.5 in
Agincourt	DB		14 × 12 in
St Vincent	DB (10)		8 × 12 in
Belerophon	DB (10)		8 × 12 in

5th Battle Squadron

Queen Elizabeth	FDB	8 × 15 in
Warspite	FDB	8 × 15 in

1st Cruiser Squadron

Warrior	AC	(6)	4 × 9.2 in
		(4)	2 × 7.5 in
Berwick	AC	(14)	9 × 6 in

4th Light Cruiser Squadron

Yarmouth	LC
Falmouth	LC

2nd and 3rd Flotillas
Each 4 D

At Cromarty
2nd Battle Squadron

King George V	DB	10 × 13.5 in
Ajax	DB	10 × 13.5 in
Thunderer	DB	10 × 13.5 in
Conqueror	DB	10 × 13.5 in

2nd Cruiser Squadron

Sutlej	AC 2 × 9.2 (10)	5 × 6 in
Bachante	AC 2 × 9.2 (10)	5 × 6 in

5th Cruiser Squadron

Dartmouth	LC
Cleopatra	LC

4th Flotilla
4 D

At Rosyth

The Battlecruiser Fleet
—Vice-Admiral Sir D. Beatty

1st Battlecruiser Squadron

Lion (Flag)	BC	8 × 13.5 in
Queen Mary	BC	8 × 13.5 in

2nd Battlecruiser Squadron

Indefatigable	BC		8 × 12 in
Inflexible	BC	(8)	6 × 12 in
Invincible	BC	(8)	6 × 12 in

3rd Cruiser Squadron

Galatea	LC
Comus	LC
Caroline	LC

6th Cruiser Squadron

Arethusa	LC
Royalist	LC

6th Flotilla
4 D

At Harwich

The Harwich Force

1st, 5th and 7th Flotillas
each 4 D

At Dover

The Channel Fleet—Vice-Admiral Heath

3rd Battle Squadron			6th Battle Squadron		
King Edward			*Lord Nelson* (Flag)	B	4 × 12 in
VII	B	4 × 12 in (4) 2 × 9.2 in		(10)	5 × 9.2 in
Africa	B	4 × 12 (4) 2 × 9.2 in	*Venerable*	B	4 × 12 in
Hibernia	B	4 × 12 (4) 2 × 9.2 in	*Bulwark*	B	4 × 12 in

Narrative

It is June 27 1916. Daylight lasts from 3 am until 9 pm.

Vice-Admiral Scheer has issued his orders covering the attack on Dover. The fleet will have steam up for 10 pm and will weigh anchor at 11 pm. The Scouting Force will leave first, followed immediately by the Battlefleet, both forces proceeding in line ahead. On emerging from the south-east swept channel cruisers will be disposed ahead and abeam of both forces at between 10,000 and 20,000 yards distance.

The Scouting Force will steam west-south-west for 120 miles, south-west for 120 miles and then turn south-south-west on course for Dover, at 25 knots. On reaching Dover the dockyards will be shelled for 30 minutes and then the force will retrace its steps and rendezvous with the Battlefleet.

The Battlefleet will follow the same course as the Scouting Force but at 18 knots, the maximum speed of its 2nd Battle Squadron. On meeting the returning battlecruisers and cruisers the fleet will turn about and escort them home. If circumstances should be favourable the 4th Cruiser Squadron will make a diversionary raid upon Great Yarmouth.

If the battlecruisers should encounter units of the enemy fleet they will, dependent upon the enemy's strength, attack and destroy them, lure them on to the Battlefleet or hold them until that force arrives in support, or, if the enemy main fleet is met, give warning to the Battlefleet and then flee for home in a direction away from the battleships.

If a suitable enemy force can be fought to destruction this may be considered success enough for one day and the Dover raid called off.

Not a brilliant plan, indeed one might even say a plan full of weaknesses, but Jellicoe, as we shall see, is not on good form either. He has had forewarning of the German raid and so is able to put to sea at 7.0 pm, a full four hours before Scheer, but the intercepted message has not been clear about the target, only stating that it will be south of Scarborough. Jellicoe has got it fixed firmly in his mind that the attack will be upon shipping in the Thames estuary and has based all his plans on that assumption.

The Grand Fleet at Scapa Flow is ordered to make a course south-east and rendezvous with the 4th Battle Squadron off Frazerburgh in the north. The Grand Fleet will then form a line abreast of divisional columns with armoured cruisers 20,000 yards ahead and light cruisers the same distance away on the beams. The whole force will steam south-south-east at 21 knots on course for

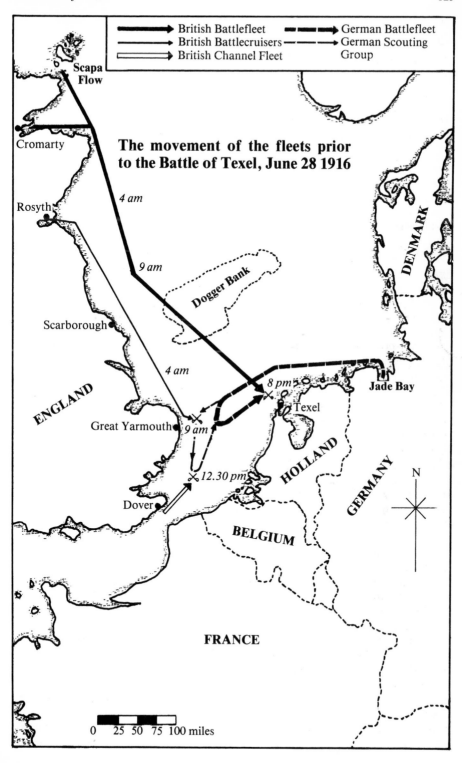

The movement of the fleets prior
to the Battle of Texel, June 28 1916

British Battlefleet
British Battlecruisers
British Channel Fleet
German Battlefleet
German Scouting
Group

Scapa Flow

Cromarty

Rosyth

4 am

9 am

Dogger Bank

Scarborough

4 am

8 pm

Jade Bay

Great Yarmouth 9 am

Texel

DENMARK

ENGLAND

HOLLAND

GERMANY

12.30 pm

Dover

N

BELGIUM

FRANCE

0   25 50 75 100 miles

the Thames (a destination which, even if it was to be the German target, could not be reached before the enemy had attacked and returned home).

The Battlecruiser Force is also ordered to the Thames, at 25 knots.

The Harwich Force is ordered to lurk off Sheerness in the estuary itself, where Jellicoe confidently expects its torpedoes to wreak havoc upon the Germans in the confines of the narrow waters.

The Channel Fleet is ordered to cruise off Ramsgate.

*   *   *

The German Scouting Force leaves harbour at 11 pm and steams west-south-west through the clear channel in the minefield, working up to 25 knots. Soon after dawn Hipper gets a shock when the *Graudenz,* 15,000 yards ahead, reports smoke dead ahead at about 20 miles distance. If the British are met at this early stage it means that all the assumptions of their dispositions on which the CinC's plan is based are wrong. The *Graudenz* is ordered to investigate and to Hipper's immense relief reports the smoke to be from a Danish liner. Hipper had not expected to meet neutrals! At 4 am the force turns south-west. At 6.30 am smoke is sighted again, by *Stralsund* on the starboard wing this time; again it turns out to be from a neutral merchantman.

At 8.55* am smoke is again reported from *Graudenz,* at 23 miles due west. The cruiser investigates while the battlecruisers steam on—Hipper has got used to sighting neutrals. Five minutes later, though, *Graudenz* signals back that she has sighted enemy light ships. Hipper racks his brains and looks at the chart— who can it be? Neither the Grand Fleet nor the Battlecruiser Force can be nearer than 100 miles, so the light ships must be the Harwich Force, perhaps supported by the Channel Fleet. He has no fear of the destroyers, and positively looks forward to meeting the Channel Fleet, because with 36 big guns against their 24, and with an 8-knot superiority he anticipates a walk-over victory. He orders a four-point turn to the west.

At 9.05 *Graudenz* identifies the ships as British cruisers and espies smoke beyond them. The enemy are on a course east-south-east by east so it looks like a head on clash, but who are they, with cruisers?

At 9.15 *Graudenz* is within gun range of the enemy cruisers (12,000 yards) and Hipper must decide whether to engage or not—the smoke beyond is confirmed as being from capital ships but whether they are pre-Dreadnoughts or not the *Graudenz*'s Captain is unable to say. Hipper suspects that they are not, that they are Beatty's battlecruisers who have stolen a march on him. Nevertheless he gives the order to engage—come what may it is his duty to discover the identity of the force and report it to the CinC with the Battlefleet.

At this stage the umpire lays the models out on the floor. The German battlecruisers are 27,000 yards (13½ nautical miles) from the British cruisers and therefore recognisable so the models are placed in position. The British big ships are 32,000 yards (16 nautical miles) from the *Graudenz* and so visible but unrecognisable. They are represented by a plain marker until they come into view.

At 9.15 the *Graudenz* and both British advanced cruisers turn four points to starboard so as to bring their broadsides to bear. They pass each other, firing, at

* Note that each of these timed sections is one move and *all* the activities in that section take place that move.

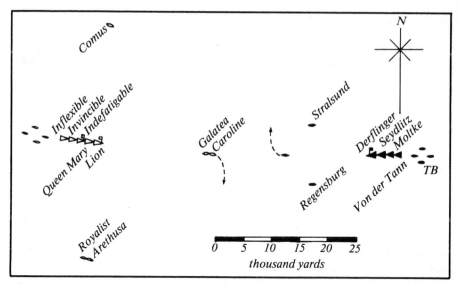

The positions of the two battlecruiser forces at laying down—9.15 am.

under 12,000 yards, *Graudenz* hitting *Caroline* with a 6 in shell and receiving two 6 in and a 4 in in reply.

By 9.20 *Graudenz* is able to identify the capital ships as battlecruisers and signals the information to *Derflinger* (the umpire replaces the plain markers with models). The *Stralsund* and *Regensburg* close on *Graudenz* to give her support. *Arethusa* and *Royalist* turn north-east to assist *Caroline,* but they are a long way off. Both *Graudenz* and *Stralsund* hit *Caroline* with a 6 in shell, whilst *Graudenz* is repaid with three 6 in and a 4 in, one of her four 6 in guns is put out of action. Meanwhile both Battlecruiser Forces are working up to maximum speeds to go to the aid of their cruisers.

At 9.25 the two battlecruiser lines are closing fast and the light cruisers turn to get out of the way. The British turn two points to port putting them on an easterly course. The lines will pass on the starboard sides so the cruisers and destroyers move towards their port, unengaged sides. The big ships are now well within range but so fine on each other's bow that no broadsides can bear.

At 9.30 the two lines pass at 12–14,000 yards range. All five British battlecruisers fire on *Derflinger* at the head of the enemy line, hitting her six times. One of the shells knocks out Y turret and another hits the steering gear, jamming the rudder to starboard. *Derflinger* and *Seydlitz* fire on *Queen Mary* hitting her twice and destroying A turret. *Moltke* and *Von der Tann* fire on *Lion* destroying her A turret and jamming *her* rudder over to starboard.

9.35: we are now treated to the ludicrous experience of having both Admirals trying to control the battle from Flagships which are circling out of control in the direction of the enemy. Beatty realises that he is going to lose the enemy unless he changes course. He can turn 16 points (ie, 180 degrees) by turning his line in succession, but this would render the turning ships liable to heavy damage and also mask the firing of the rearward ships. He decides upon two successive eight-point turns by squadron which, will put *Indefatigable* in the lead and *Queen Mary* at the rear. If he turns to port he will lengthen or lose the range, which he does not want, so the turn is ordered *towards* the enemy. The

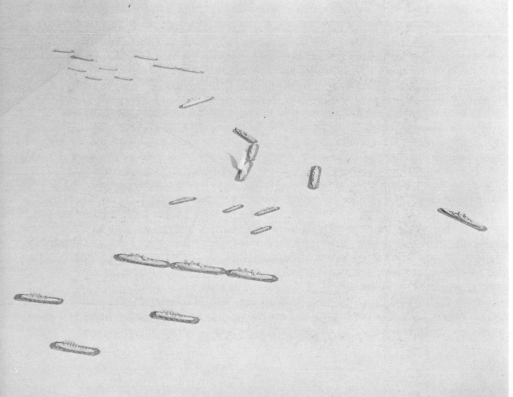

Above *The battlecruisers open fire.*

Below *As the two Flagships circle out of control the* Indefatigable *is heavily damaged by German gunfire.*

light cruisers and destroyers are simply ordered to turn 16 points as they can manage it in one operation.

Hipper is suffering from the British 13.5 and 12 in guns—his are mostly 11 in—so he decides upon a torpedo attack. He flies a signal for his destroyers to cut through the line and attack the enemy.

Both Admirals' orders will become effective next move. Meanwhile the two circling Flagships suffer as the enemy find their range (treated as if turning in succession) *Lion* is hit in the hull by four shells from *Seydlitz,* and *Derflinger* is fired at by all the British ships except *Lion* (who cannot fire while circling) and *Invincible* who is masked by *Lion* and suffers eight hits, one of which puts A turret out of action. *Moltke* and *Von der Tann* fire on *Indefatigable* and hit her three times, destroying B turret.

By 9.40 the engineers have not yet succeeded in unjamming *Lion*'s rudder (ie, the dice throw is too low) so she continues to circle. The other ships of the British Fleet, though, obey the signals she is flying; the 1st Battlecruiser Squadron (Queen Mary alone) and the 2nd Battlecruiser Squadron turn eight points to starboard independently and steam towards the enemy in parallel lines, whilst the destroyers and cruisers immediately go about by 180 degrees and open up a large gap between them and the battlecruisers.

Beatty realises his mistake in allowing the light ships to become separated from him when he sees the enemy torpedo boat flotilla come surging through the gaps between the battlecruisers. His destroyers, which should be protecting him from such onslaught, are steaming off, eight miles away.

The German battlecruisers continue on their westerly course, firing at the *Indefatigable,* who, leading the 2nd Battlecruiser Squadron line, is closest to them, at less than 8,000 yards. Nine shells smash into her, destroying her A turret, bringing the director tower down, and bringing her past the 50 per cent damage mark so that she loses speed. The three ships of 2nd Battlecruiser Squadron fire on *Seydlitz,* knocking out Q turret and hitting her twice in the hull. *Queen Mary* fires on *Von der Tann* but the range is fogged by torpedo boats and the fire controllers are off-put by the 90-degree turn and she only scores one hit. She also fires her 4 in guns at the attacking destroyers and scores three hits; the ships of the 2nd Battlecruiser Squadron cannot bring their secondary guns to bear on the torpedo boats which are dead ahead of them.

At the end of the move three of the German torpedo boats launch full four-torpedo salvoes at the *Queen Mary.*

By 9.45 the British battlecruisers should be carrying out their second eight-point turn to starboard this move, but as on this course they find themselves more or less combing the torpedoes they dare not turn, so that instead of running parallel to the enemy they cross his rear at right angles.

All 12 torpedoes skim dangerously near *Queen Mary,* but to Beatty's intense relief, they all miss. Half-way through the move the torpedo boat *V2* sneaks up close to *Indefatigable* and launches the four torpedoes which she has been holding back.

Derflinger's engineers manage to free her rudder, and so she turns and steams off at maximum speed to try and catch her fellows, who are all firing on *Inflexible* who, being at the end of the 2nd Battlecruiser Squadron line is the only one on which broadsides can bear. Five shells smash into her hull, a sixth destroys B turret and a seventh bursts deep in her magazine. A momentary delay (whilst a dice is thrown—2!) and she erupts in a column of flame which splits

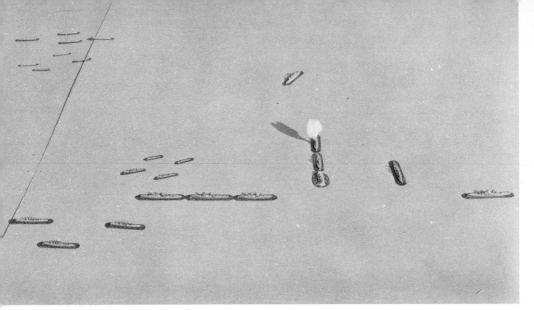

Indefatigable is sunk by a torpedo and Inflexible blows up. In the distance are the British cruisers and destroyers who should have been supporting the big ships.

her in two. To complete the disaster one of the torpedoes from *V2* strikes *Indefatigable* amidships, explodes, and causes great damage (28 points) which in *Indefatigable*'s battered state is sufficient to sink her. The only damage inflicted on the Germans is one turret destroyed, and a hulling hit on the *Von der Tann* at the tail of the line.

It is now 9.50 and Beatty has had enough. With his Flagship still out of control he has only two battlecruisers still able to fight and one of these is of a class which has shown itself too fragile for such a battle. He signals a general recall and without battlecruisers to support them he dare not even risk his destroyers in a delaying attack on the enemy line. The two fleets lose sight of each other and are returned to the map.

Hipper steams on westward until out of sight of Beatty. The *Derflinger*'s speed has been reduced to 23 knots and the *Graudenz* is also badly damaged, so Hipper transfers his flag to *Seydlitz* and sends the two battered ships home. The torpedo boats have used up all their torpedoes so they are also ordered home. Cock-a-hoop with his success, Hipper is determined to continue with the raid and so orders his remaining five ships on a course south-south-west for Dover.

Beatty continues to circle out of control. When, after half an hour, the *Lion*'s rudder is eventually freed the Vice-Admiral is so despondent that he forgets his scouting duty completely and orders a return to base.

*   *   *

At 9.15 when Jellicoe receives Beatty's report of sighting the enemy battle-cruisers he does a little belated thinking and, looking at his map, realises at last that he is not going to reach the Thames (which he still assumes to be the target) in time. Consequently he changes his plan and determines upon cutting the enemy off from home, so he orders a change of course to the south-east heading for the entrance to the Jade Bay. At the same time he issues orders to the Channel Fleet to steam north in support of the battlecruisers—no orders were issued to the Harwich Force but he strenuously denies the vile slander that he had forgotten its existence. The destroyers continue to lurk menacingly in the Thames estuary.

Hipper continues to steam on south-south-west with his three battlecruisers and with *Stralsund* and *Regensburg* scouting ahead. At 11.30 smoke is sighted which he hopes is the Channel Fleet. His brush with the battlecruisers has only served to whet his appetite for combat. The sighting is only of another neutral, though. At 12.15 smoke is again sighted south-south-west by west. The *Stralsund* investigates and at 12.25 is able to report that the Channel Fleet is ahead. Hipper orders full steam ahead, confident of a second success. His force is now, though, much reduced in strength; *Derflinger* has gone home and both *Seydlitz* and *Von der Tann* have lost a turret reducing his broadside of big guns to 24 11 in compared with the enemy's 24 12 in.

At 12.30 the models are laid down by the umpire and the battle comences.

By about 12.30 or 12.35 Heath is not at all confident of his chances of defeating the enemy, but what he knows he must do is to prevent them passing, and to inflict sufficient damage on them to force them to return home. He decides to take the risky step of dividing his forces, taking the 6th Battle Squadron himself across the enemy's line of approach and sending the 3rd Battle Squadron to block them from the East Coast. He orders the 6th Battle Squadron to turn east at 18 knots and the 3rd Battle Squadron to continue on course north-east.

Hipper is well pleased with the British separation—he plans to drive in

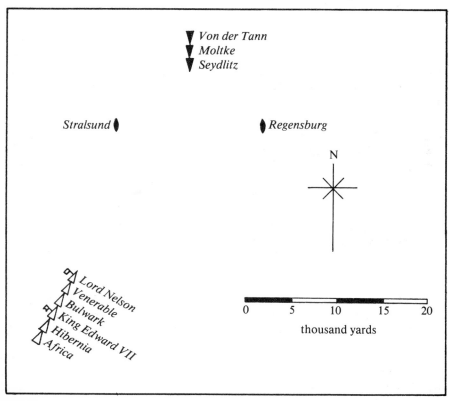

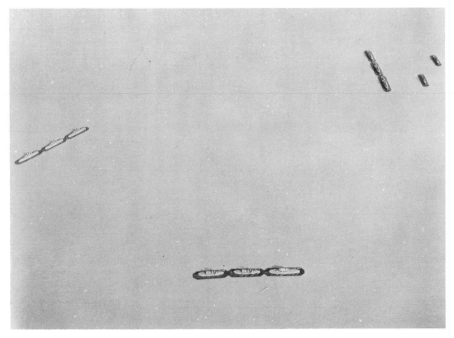

The pre-Dreadnought battleships of the Channel Fleet decide to block Hipper's advance.

between them and then use his superior speed to destroy them separately. He orders his cruisers to fall back to his port wing.

At 12.40 the Germans steam on at 26 knots and fire on the 3rd Battle Squadron at 16,000 yards hitting *King Edward VII* once in the hull. Both British squadrons return fire on *Seydlitz, King Edward VII* destroys her B turret and *Venerable* hits her in the hull.

By 12.45 the 3rd Battle Squadron has now closed to less than 12,000 yards and the 6th Battle Squadron is almost crossing the Germans' T at about 6,000 yards range; they are blazing away with 12 in and 9.2 in guns and straddle *Seydlitz* several times, hitting her with nine 12 in and two 9.2 in shells and putting out of action two of her three remaining turrets—the supposed 'old crocks' are showing their teeth and Hipper realises his mistake. The *Seydlitz* cannot last long at this rate and she is now considerably out gunned—he hoists a signal to retire.

Meanwhile the *Von der Tann, Moltke* and *Seydlitz* are still firing well and manage to hit *King Edward VII* eight times.

At 12.50 Hipper's signal becomes effective and the squadron turns around 180 degrees in a *Gefechtskertwendung* to starboard and they steam north, away from the battle, the raid abandoned. The two squadrons keep up their fire on the *Seydlitz* and hit her eight times, knocking out her last turret and reducing her to nine knots

Just as the British become elated at driving off the enemy, *King Edward VII* is straddled again. She is hit in the magazine and after a tense moment she explodes amidships and breaks in two.

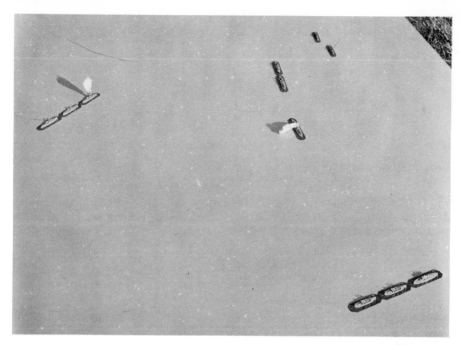

As the German battlecruisers retreat and leave the damaged Seydlitz *to her fate, a salvo hits* King Edward VII *and she explodes.*

At 12.55 *Von der Tann* and *Moltke* speed on and abandon the crippled *Seydlitz* (with Hipper) to her fate. The old battleships are too slow to give chase, so they close in for the kill on *Seydlitz*, the *Hibernia* and *Africa* steaming on past the wreckage of their Flagship and the 6th Battle Squadron turning north-east. *Seydlitz* takes eight more hits and starts to settle by the stern, dead in the water. *Hibernia* takes three hits by way of a Parthian shot from the retreating battlecruisers and leads *Africa* in to pick up survivors.

The forces move out of sight of each other and are 'returned' to the map.

The *Moltke* and *Von der Tann* steam northwards and at 3 pm, 45 miles off Lowestoft, they rendezvous with the Battlefleet which turns about and sets off homeward. Scheer rightly suspects that the British forces are much nearer than he had originally anticipated so instead of retracing its steps the High Seas Fleet follows a course close to, but out of sight of, the Dutch coast. Meanwhile the Grand Fleet is on course for the entrance to the Jade Bay and it is a question of who will get there first.

As 8 pm approaches Scheer becomes more confident of having eluded Jellicoe, but shortly afterwards the *Emden,* scouting 10,000 yards ahead, reports smoke to the north which ten minutes later is identified as the *Bachante* Class armoured cruisers. They can only be an advance force of the Grand Fleet, and Scheer looks at his map; he is trapped against the Dutch coast; whichever way he goes he must inevitably be forced into action with the overwhelming force of British Dreadnoughts. His only hope is to continue on course and hope that the enemy is so far away that they will still not have stopped the route home when darkness falls in an hour's time. The Battlefleet maintains its course, therefore,

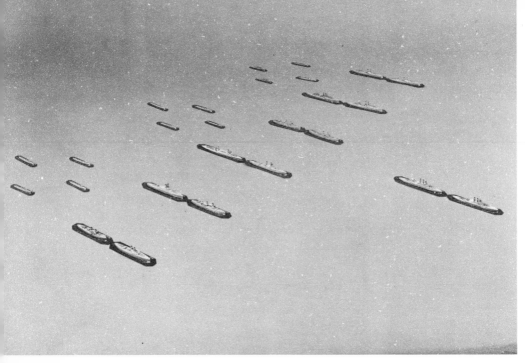

Above *The Grand Fleet, with the 5th Battle Squadron in the van, steams on at full speed to cut the enemy off from his base. Out of the photograph there are armoured cruisers ahead of the fleet and light cruisers on the wings scouting for the High Seas Fleet.*

Below *The High Seas Fleet steams for home, with the two remaining battlecruisers and light cruisers of Hipper's Scouting Group bringing up the rear. Ahead and out of sight are the light cruisers.*

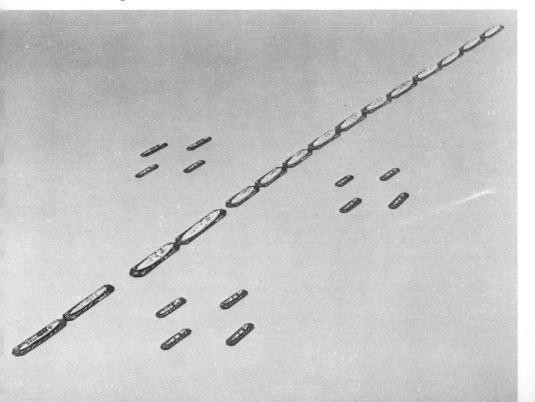

and still at 18 knots! Scheer begins to regret bringing his pre-Dreadnoughts with him as they are of little fighting value and their low speed is a serious handicap on his run for home. He is determined not to abandon them, though. At 8 pm the *Bachante* and *Sutlej* open fire on the German light cruisers at 16,000 yards range and the last stage in the battle commences.

The umpire lays the models out on the floor. The Dreadnoughts of the Grand Fleet are still out of recognition distance of the German cruisers but with only an hour of daylight remaining Jellicoe is impatient of delay and eschews the use of plain markers.

At 8.0 pm Jellicoe sees that he has the enemy in a trap, but can he spring it before nightfall? A deployment on his starboard column is the quickest way of getting into big gun range but this would result in the Germans crossing his T and not being cut off. He deploys, therefore on his port column on course south-east whilst ordering the fast battleships of the 5th Battle Squadron to steam ahead at full speed and attempt to turn to the head of the enemy line. The destroyers and cruisers move forward to cover the ponderous evolutions of the battle squadrons from interference from enemy torpedo boats.

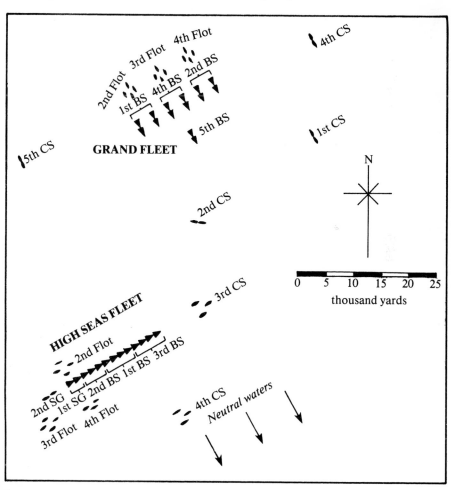

The armoured cruisers of the 2nd Cruiser Squadron, *Sutlej* and *Bachante,* open a heavy fire upon the *Emden* leading the German Cruiser Squadron ahead of their Battlefleet, damaging her badly.

At 8.05 the German Fleet continues on course north-east whilst the Grand Fleet continues its deployment into line ahead on a converging right-angle course. Both sides need to reach the intersection point first, but with the two *Queen Elizabeths* steaming on at 25 knots against the Germans' 18 there is little doubt who will win. In the open sea the Germans would turn a couple of points to starboard but with neutral waters to that side they have no option but to maintain that course.

The German 2nd and 3rd Flotillas steam out to attack the deploying Grand Fleet, just as Jellicoe anticipated, whilst the 4th Flotilla moves ahead to engage the two armoured cruisers whose fire has just turned *Emden* into a blazing wreck.

At 8.10 both fleets continue on course. The two armoured cruisers shift their fire from the now sinking *Emden* to the *Leipzig* and *Dresden* behind her, but now the leading German Dreadnought, *Kaiser,* has found their range and hit *Sutlej* with two 12 in shells. The sally by the German Flotillas, supported by the 4th Cruiser Squadron is countered by the British destroyers and cruisers steaming down to engage them.

By 8.15 both the German light cruisers and the British armoured cruisers find the fight too hot and turn away from each other; this does not prevent the *Sutlej* receiving another two 12-inchers from the *Kaiser,* damaging her badly. The main fleets are still well out of range of each other but the *Queen Elizabeth* and *Warspite* are closing fast on the intersection point.

On the right the Grand Fleet deploys into line ahead. On the left is the German Fleet with its cruiser, Emden, *suffering in her fight with the British armoured cruisers.*

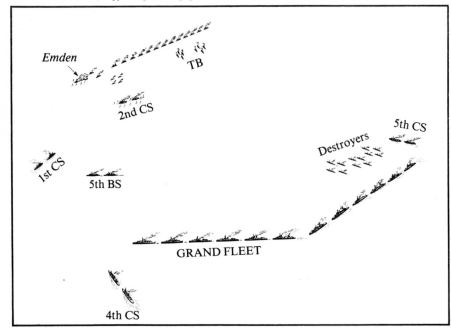

At 8.20 *Kaisers'* shells hit *Sutlej* yet again—she is reduced to 3 knots and makes an easy target for one of the boats of the 4th Flotilla which launches a full salvo of four torpedoes at her. The other three boats, though, ignore the armoured cruisers and speed on towards what they now see is the real danger—the 5th Battle Squadron. These two super-Dreadnoughts have opened fire at extreme range and a 15 in shell from one of them smashes into the *Kaiser*. The two destroyer forces come within range of each other and a rapid fire commences as they approach at a combined speed of 60 knots.

At 8.25 the German line is being headed and the fact is brought rudely to Scheer's attention by the explosion of three 15 in shells in *Kaiser*'s hull. Scheer is at a loss as to what he should do, though; he cannot turn away south, a turn north attempting to cut behind the British would subject him to fire from the whole length of the Grand Fleet and a turn in succession within range of the enemy, and a turn away south-west would take him directly away from the safety of home. Unable to decide, he does nothing, whilst getting a small crumb of comfort from seeing the *Sutlej* rent apart by a torpedo explosion.

Meanwhile the two forces of destroyers, torpedo boats and cruisers swirl into a scrambling mêlée between the two battle lines. The fight is hot and the heavier gun armament of the British ships begins to tell.

At 8.30 the boats of the German 4th Flotilla launch their torpedoes at the 5th Battle Squadron and turn away. Those two ships continue to fire on *Kaiser,* who is now suffering badly and is starting to lose speed.

In the destroyer mêlée the cruisers are inflicting heavy damage; *Dartmouth* and *Cleopatra* each sinking a torpedo boat whilst the destroyers *Umpire* and *Undine* succumb to the broadsides of the *Breslau* and *Nürnburg*. The *Nürnburg,* however, is seriously damaged in her turn by the 9.2 and 6 in guns of *Bachante.*

By 8.35 destroyers and torpedo boats are sinking and lying dead in the water but the firing continues as others steam about at full speed. The 1st Cruiser Squadron, *Warrior* and *Berwick,* plunge into the fray and are joined by *Bachante* but the three armoured cruisers have wandered too close to the German battle line and are repeatedly straddled and hit by heavy shells.

The *Queen Elizabeth* and *Warspite* steam on, disdaining to turn from the torpedoes, which fortunately run by them without hitting. They continue to fire on *Kaiser,* whose remaining forward turret at last gets a hit back on *Queen Elizabeth*—the effect of a 12 in shell, though, is puny compared with the smashing power of a 15-incher. The van of the Grand Fleet is now within range of the leading German ships and *König, Grosser Kurfürst* and Scheer's own ship, *Friedrich der Grosse,* and *King George V, Ajax, Thunderer* and *Conqueror* engage each other. Shells of 12 in calibre are not fair exchange for 13.5-inchers and the Germans begin to suffer.

It is 8.40 and the 5th Battle Squadron has now passed the intersection point and it is obvious even to Scheer that the remainder of the Grand Fleet will be crossing his T in a few minutes. The full horror of the situation moves Scheer to action and at last he issues an order to his fleet.

Meanwhile in the destroyer mêlée the heavy firepower of the British ships is overwhelming and the German Flotilla Commander orders a withdrawal. The already badly damaged *Nürnburg* is hit by a torpedo from the destroyer *Mameluke* and sinks.

The armoured cruisers have become targets for most of the German Battle-

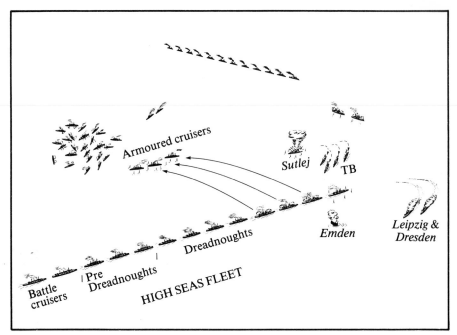

Above *The* Queen Elizabeth *and* Warspite *bring fire down on the German van. In the distance the cruisers and destroyers of both sides fight in a furious* mêlée.

Below *The Grand Fleet crosses the T. The German cruisers and torpedo boats withdraw from the* mêlée.

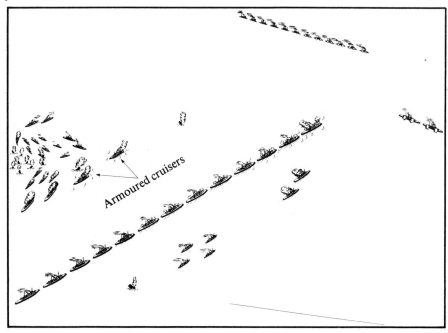

fleet and they suffer for their mistake of going too close—*Bachante* is hit six times and capsizes, whilst both *Warrior* and *Berwick* are badly damaged.

The 5th Battle Squadron and leading ships of the British line, who are now crossing Sheer's T pour a heavy and accurate fire into the German van at less than 12,000 yards range, whilst the Germans can only reply with their full broadsides at long range on the British centre. The *Kaiser* is now a blazing wreck and reduced to 6 knots.

At 8.45 Scheer's order becomes effective and the whole line of battle swings about in a *Gefechtskertwendung* to port, all, that is, but the *Kaiser* who is so badly damaged that she can only manage 6 knots and falls behind. Jellicoe is surprised by the German move and continues on course. The 5th Battle Squadron's role, though, allows its Vice-Admiral to turn and follow at the end of the move without orders from the CinC.

The crippled *Kaiser* is repeatedly hit, not only by the 5th Battle Squadron but also by the armoured cruisers who are attempting, with *Dartmouth* and *Cleopatra,* to keep close to the enemy for when darkness falls. The battering is too much and *Kaiser* starts to go under. The Dreadnoughts of the German rear and the British van exchange fire but the range has now lengthened and accuracy is poor. The other battleships of the German line continue to fire on the British armoured cruisers and cruisers hoping to drive them off before darkness falls.

At 8.50 Jellicoe hoists a signal to his Battlefleet to turn towards the retreating enemy but until the signal is acknowledged it cannot become effective. The enemy are now out of gun range and the only British force still in contact with them is the 5th Battle Squadron, the cruisers and armoured cruisers having been

The Germans retreat. The Grand Fleet turns by division and pursues, but night falls.

5th BS

compelled to withdraw out of range and the destroyers having pulled back ten minutes ago to lick their wounds.

Queen Elizabeth has her A turret damaged by a shell from *Grosser Kurfürst*, and *Warspite* is hit by *König*. The German ships are also hit but none is slowed down.

At 8.55 Jellicoe's last order becomes executive and the Grand Fleet turns eight points to port by division. It is too late though; only the *Queen Elizabeth*s are still in range of the Germans and in five minutes it will be dark. Despite that the 5th Battle Squadron puts in some good shooting and *Grosser Kurfürst* is reduced to 17 knots.

It is 9.00 and darkness falls. The enemy fleets are out of sight of each other and continue on their courses for a few minutes and then Scheer orders an eight-point turn in succession. This eventually takes him into neutral Dutch waters but he is less concerned at the moment by International Law than by the fear of running into the Grand Fleet again.

The forces are returned to the map and the umpire plots their movements in the dark. The British forces completely lose contact with the Germans and by midnight Scheer is dropping anchor in the Jade Bay whilst Jellicoe, although beginning to despair of remaking contact, continues to steam off Holland looking for him.

At the end of the day the relative losses are as follows:

	BRITISH		GERMAN	
	Sunk	**Badly damaged**	**Sunk**	**Badly damaged**
Dreadnoughts	—	1	1	3
Battlecruisers	2	1	1	1
Pre-Dreadnoughts	1	—	—	—
Armoured cruisers	2	2	—	—
Light cruisers	—	3	2	4
Destroyers	3	4	4	7

The Germans have obviously inflicted more damage on armoured ships than they have suffered, five sunk to two, but three of the British ships were obsolete types of little fighting value. Scheer failed to achieve his object, and his lucky escape off Texel and subsequent scurry for home will probably have had a demoralising effect both on his Fleet and on the German public.

The umpire therefore declared a victory for the British, but without honours on either side.

Chapter 12

Other periods

If we are going to be over simplistic we can say that sea fighting can be divided into three distinct historical eras, the age of ramming and boarding, the age of gunpowder and sail, and the age of armoured steamships. All periods of naval warfare, you will find, can be categorised into one of these classes, or into a combination of two where technical development means that two periods overlap. I have given rules for the most popular periods within each of these three 'ages' and so it naturally follows that with a few alterations and amendments, to take account of changes in weapons and fighting techniques, the rules in this book can be used to fight wargames of any period in history.

The Middle Ages

Mediaeval sea warfare in the Mediterranean is easy to re-fight because the galley still reigned supreme and fighting techniques had changed very little since ancient times. The rules in Chapter 4 can be used almost unchanged for this period. The only difference being that there was more emphasis on boarding than on ramming.

In northern waters sea battles were contests between clumsy, unmanoeuvrable sailing ships with boarding their only means of attack. Such sea battles were truly land battles at sea and it was not uncommon for the defenders to chain their vessels together to form a stationary floating battlefield. I have refought the Battle of Sluys once, and fought out three other naval battles in this era but they were all slogging matches with little tactical interest.

The Renaissance

In the Mediterranean galleys continued to rule the sea—larger than their ancient forbears and mounting a handful of guns in the bows but largely unchanged in their fighting techniques. The cannon were slow-firing and clumsy, and only really used to keep the enemy annoyed whilst the fleets closed for the serious business of boarding. The galley rules in Chapter 4 can easily be modified to cover this period—all that is necessary is to introduce cannon rules (the rules regarding stone-throwing engines will do) and to allow the new lateen-rigged ships to sail to within 45 degrees of the wind. Many of the crewmen would be armed with firearms but the different effects of arquebus fire and archery need not worry us unduly.

In northern waters the clumsy sailing merchant ships were evolving into proper sailing battleships with a few pieces of heavy artillery poking through

ports in the hull side, but with most of the armament consisting of small man-killing hand guns firing down from the massive castles at the fore and aft of the ship. These ships were still horribly bad sailors, being built as floating impregnable castles rather than mobile, manoeuvrable ships of attack, and so were augmented in most navies by galleys and galleasses, this latter a cross between the heavy, broadside-armed sailing ship and the manoeuvrable, swift galley. An excellent wargame can be had by combining a few Great Ships, Great Galleys, galleasses and galleys, with all their contrasting qualities, into a couple of fleets of the early 15th century. The system of rules given for Napoleonic warfare can be used, with additional rules relating to oar-propelled vessels. The following characteristics should be catered for in the rules.

1 The Great Ships will sail slowly, and very badly to windward. Allow them to sail no closer than 60 degrees to the wind.

2 Rowing ships such as galleys and galleasses will be able to sail at least as well as the Great Ships and when they are lateen-rigged will be able to sail to within 45 degrees of the wind.

3 Boarding was still regarded as the proper way of fighting at sea and so large numbers of soldiers were carried. Make boarding factors much larger than in the Napoleonic period and give advantages in boarding fights to high-sided ships against low ones.

4 Give ships two firepower factors. One represents the heavy artillery which can damage the hull, masts, guns and kill crewmen, the other is only used at very short range and represents the fire of the man-killing swivel and hand guns in the castles. This fire is only effective against men and so reduces the boarding factors without affecting the rest of the ship.

The Armada

The Spanish ships which formed the Armada were of the Great Ship type, as described above, with large complements of soldiers for boarding, shortish-ranged heavy cannon in the hulls and man-killing light guns in the high castles. The Spaniards also employed galleasses, and would also have included galleys, had these not been forced to return home by bad weather. The English had more or less abandoned rowing ships by this time, and only a handful of Great Ships remained. Most of the English warships were low, race-built galleons which combined speed and the ability to sail quite close to the wind with an armament of long-range medium guns firing shot of 16 or 18 lb weight.

1 Allow English ships to sail within 45 degrees of the wind but Spaniards only within 60 degrees. English ships should be moved at frigate speed and Spaniards as ships of the line.

2 Spanish ships will have two fire factors, one for man-killing guns and one for artillery. The artillery factor will generally be as heavy or heavier than for English galleons but the maximum range will be less, say two thirds of English culverin range. English ships will be regarded as having no man-killing guns (because they have no castles in which to place them).

3 The boarding factors of Spanish ships will be four or five times greater than for the equivalent-sized English ship.

Above left *A mediaeval cog, manned by 20 mm scale knights and archers. Her flag is that of the Cinque Ports.*
Left *Great Ships, galeases and galleys engage as Henry VIII's navy sallies out to prevent a French landing on the South Coast.*

The Dutch Wars and the 18th Century

Galleys continued in use in the Mediterranean, large rowing squadrons being present at both the Battles of Malaga (1704) and Matapan (1717).

In the Baltic, galleys, row barges and oar-propelled gun launches were used in very large numbers right up to the Napoleonic Wars by both the Russian and Swedish Navies. As these fleets fought mostly amongst the shoals and islands of the Finnish Archipelago the rowing vessels had some real advantage over the sailing ships, and so although the Russo-Swedish Wars are only rarely refought by wargamers they can provide some absolutely fascinating sea battle games.

By the time of the Dutch Wars sailing ships had evolved into ships of the line little different from those which fought at Trafalgar. The ships were generally smaller, gun for gun, than their later equivalents with the result that gun ports were nearer the water and lee side lower deck guns unable to fire when it blew 'more than a cap full of wind'. This latter can have important tactical consequences so it is worthwhile reproducing it in our rules. In any battle before 1750, therefore, when the wind is fresh or stronger, reduce the lee side fire effect of three-deckers by a third and of two-deckers by a half.

A curious feature of naval warfare in this period is the number of post-battle accusations of subordinates failing to support their commanders. Blake after the Ness, King William after Beachy Head, Benbow after his single-handed battle in the West Indies, Mathews after Toulon and Howe after the Glorious First of June, all complained of portions of the fleet hanging back and not coming to grips with the enemy. There are two possible reasons for this problem, firstly the navy did not achieve a full, non-political professionalism until the Napoleonic Wars, and secondly the lack of a satisfactory signalling system before about 1790 allowed pusillanimous Captains and Admirals to misunderstand aggressive orders which had not been specified in council of war. Faint-heartedness in subordinates can be infuriating in a wargame, but if you want to allow for this happening here is a rule which we have employed once or twice with exasperating success:

1 At the beginning of the battle both commanders write down their plan of attack. This is assumed to be covered by the orders given in council of war, and no Captain or Admiral has any excuse for not carrying them out to the letter.

2 A list of orders, with corresponding signal numbers, is written. As soon as an order is signalled which in any way requires a deviation from the original plan the units of the fleet are subject to 'misunderstanding'.

3 Throw a dice for each unit of the fleet affected, ie, each ship, or Flagship if the fleet is divided into squadrons, 1 indicates a faint-heartedness, 2, 3, 4, 5, 6 indicate that the ship or squadron obeys the spirit of the order. If a 1 is thrown, throw a second dice. If you get a 6 the ship or squadron will fight, but will not break the enemy line. A 4 or 5 mean that the ship or squadron will not approach the enemy closer than medium range. A score of 1, 2 or 3 and the ship or squadron will fire at the enemy at long range and will go no closer.

Something which we have not yet tried but which may add interest to an 18th century campaign is to give all Flag Officers a personality and from their known characteristics determine the likelihood of their co-operating with the Commander in Chief. Strong Whigs and die-hard Tories were notoriously bad mixtures in an 18th century fleet.

The Russo-Japanese War

Both the Ironclad and the Dreadnought rules can be used for fighting sea battles of this period. If the 'analytical' Ironclad period rules are being used all that is necessary is to lengthen the ranges and to take account of the penetrative powers of the more modern guns and convert thicknesses of the newer types of armour to inches of iron. The reprint of *Jane's* for 1905 will give all the necessary information. If the Dreadnought period rules are used the maximum ranges of big guns will have to be reduced as gunnery control was not very highly developed at this time. The pre-Dreadnought battleships were also designed to use their 6 in batteries offensively instead of just defensively as in Dreadnoughts, so the rules must cater for the firing of small guns against armoured ships.

The Second World War

Conventional actions of the Second World War such as the Battle of the River Plate, Narvik, etc, can be fought quite easily as a wargame using the First World War Dreadnought rules, but if your interest is in the great carrier battles of the Pacific you have got problems. I have tried wargames with aircraft carriers in which card markers were used to represent formations of aeroplanes—on the underside of the card, hidden from the view of the opponent, were written aircraft type, numbers and height. The markers were moved at appropriate speed and on reaching the target they took anti-aircraft fire from the surrounding ships, the numbers of aircraft were reduced by those destroyed and then torpedo markers were launched or bombs diced for. The problem was that both fleets were at all times in view and so there were none of the difficulties of the real thing caused by fighting an enemy who was way out of sight over the horizon. The very long ranges at which carrier battles were fought also mean that wargaming with models is unsatisfactory because the smallest scale available is still far too big. I would recommend that anyone wanting to refight the great aircraft carrier battles of the Pacific war dispenses with model ships altogether and manoeuvres the fleets by erasing and re-drawing their position in pencil on a large-scale map.

The American Civil War

The naval side of the American Civil War is fascinating, if only because it was the first war in which a submarine sank an enemy warship and the first war in which Ironclads were used. Most of the warships in use were steam-propelled wooden ships but Ironclads were introduced in increasing numbers towards the end of the war. One of the difficulties of naval wargaming in this period is that gun technology had been temporarily overtaken by armour so that Ironclads had almost total invulnerability to gunfire. As a result actions between two Ironclad forces tended to result in draws, whilst actions between Ironclads and wooden ships were a foregone conclusion in favour of the former. The submarines referred to were crude manually-propelled Spar torpedo craft which ran awash for secrecy of approach and consequently they have no place in the conventional set-piece wargame. The only real way to reconstruct American Civil War sea fights, then, is as campaigns which are so devised as to allow wooden ships to make full use of the advantages they possessed over Ironclads, ie, seaworthiness, range and endurance, and in which the opportunities exist for secret torpedo attacks on moored ships.

Chapter 13

Campaigns

I have mentioned earlier in this book that, in my opinion at least, sea battle games are best played in the larger context of a campaign, so that there is a history to give the battle a meaning and purpose and a follow-up in which the real strategic consequences of the battle can develop. But properly organised campaigns in which the armchair Admiral can practise his grand strategy are also worthwhile and enjoyable wargaming in their own right. Quite what form your campaigns take will depend very much on your interest and resources and on your own interpretation of history, so I shall make no attempt in this chapter to lay down hard and fast systems of play and rules. Rather it is my intent to list some methods and campaign situations which have given me good results in the past in the hope that they will give you some ideas for use in your own operations.

Maps

A map or, more usually, a number of maps, are an essential requirement in any campaign. The sort of map generally available, eg, the atlas page, is invariably unsuitable for our purposes being usually too small in scale, covered with unnecessary and confusing detail, showing too much land and too little sea and being frequently too nice to draw over. We have therefore to make our own maps, and so I describe here a method by which even the clumsiest of us can reproduce charts and coastlines accurately, enlarging and reducing them as necessary.

The map which we are making is to be drawn on graph paper. The 1 in squares facilitate copying and on the finished map provide a reference grid. A map of the area required is found in an atlas (it does not really matter how small the scale) and covered in tracing paper. The area of the map to be reproduced is broken down into the same number of squares as there are on the graph paper, so that if the graph paper has 22 × 15 1 in squares, the small map in the atlas will be covered with 22 × 15 squares of, say, $\frac{1}{4}$ in size. The details from the small map are then copied square for square on the graph paper, so that the map is accurately reproduced, in this case four times larger.

Secret movement

Armies, particularly those of Napoleonic times and earlier, moved very slowly and, usually, through thickly populated areas, so that news of their location, movements and approximate composition raced ahead of them and

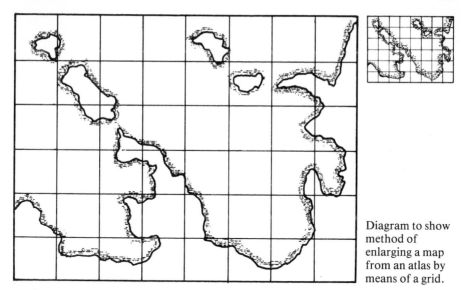

Diagram to show
method of
enlarging a map
from an atlas by
means of a grid.

every commander had a pretty good idea of where his opponents were and what they were up to. In land campaigns, therefore, it is quite realistic to use the simple system of both players pinning flags representing armies into a single map, moving them about openly and setting up a battle when they meet.

In naval warfare, though, since earliest times ships and fleets have been able to move so swiftly over great distances and without garrulous civilians blabbing about their activities, that their movements have been unknown to their enemies. To re-create these strategic operations as wargames campaigns we have to use a more complicated system of map movement which will allow both sides to move secretly but know when their forces sight an enemy.

The simplest method is to use one map with numbered flags or counters representing the various ships or squadrons of each side, but with additional markers for each side representing bogus forces. Opponents are not told what the flags represent so that until two markers come within sighting distance of each other it is not known whether the enemy's flag represents a collier, or a full battle squadron with ancillary cruisers and destroyers, or nothing at all. We do insist, though, that when two markers 'sight' each other and one of them is a bogus force its owner announces that it is a dummy so that the other player can keep the identity of his force secret; otherwise bogus markers become additional reconnaissance units.

An inevitable feature of a campaign using the bogus marker method is that the use of real and dummy flags becomes an art in itself, where concealment is achieved by a game of bluff and double-bluff with bogus markers moving purposefully and significantly around some vulnerable area so as to draw the enemy away from the real objective whilst the real forces, acting like bogus ones, spring on the target unopposed. This can often be a good game but it does not have much parallel in real naval history and so cannot be accounted a very realistic simulation of war.

A better method but one requiring more preparation (and being slower in operation) is the well-known matchbox method. Each player has a gridded map and on it he marks the movements of his own forces. In addition a number of matchboxes are glued together to form a sort of chest of drawers, each grid square on the map being represented by a matchbox in the chest. As a force moves from one grid square to another a counter representing it is moved from one matchbox to the next. When a counter enters a matchbox containing an enemy marker a sighting has been made.

There are a couple of anomalies which can occur with this system but neither is insuperable. The first is that two enemies can be only half a mile apart but if they are divided by a grid line they will be invisible to each other. We argue this away by blaming it on fog or sea mist. The second is exactly the opposite—too great a range of vision. A ship's sighting range in the matchbox method can be taken as about half the side of one of the grid squares, which, if the scale of the map is such that a grid is 200 miles square gives a visibility of about 100 miles. As the absolute distance at which the smoke from even the filthiest coal-burner could be seen would be about 20 to 25 miles it means that raiders or other forces relying on stealth and secrecy have a very much reduced chance of success. We can partially overcome this by subdividing the grid boxes into four or nine

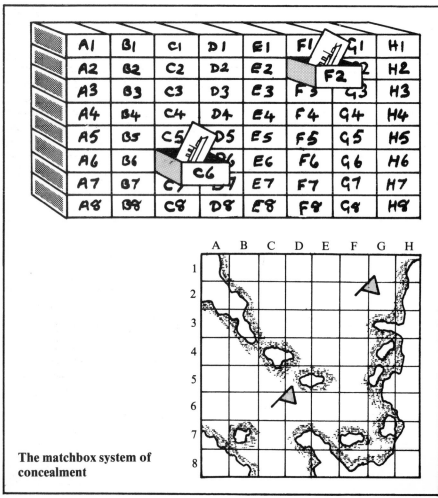

The matchbox system of concealment

smaller boxes and when forces meet in a match box announcing the sub-squares in which they are. If they are in the same one they meet and fight and if they are in different ones they go their separate ways, knowing the enemy is near but not knowing his type or composition, as though smoke is seen disappearing over the horizon. Alternately we can just throw a dice to see if they sight each other, adjusting the odds of sighting to match the amount that the grid square is larger than the proper sighting zone.

Naval warfare often involves blockade, where the forces of the inferior power find themselves most of the time confined to harbour, only occasionally sailing out on raids, blockade-running and hit-and-run operations. Blockading was a tedious business in real life for the enforcing navy and it can be a tedious business too in a wargame if hours and hours are spent moving patrols through the matchbox chest looking for an enemy who usually is not at sea. We have made this type of campaign much faster and more enjoyable by doing away with the matchboxes and hourly or daily map moves and instead plotting on the maps paths for a whole week's movement, the lines broken down by days. Grid squares on the map are grouped together into named or numbered zones (around British Isles we use the weather forecast sea areas, Rockall, Mallin, German Bight, etc). At the end of each 'map week' the player with the superior force calls out those zones which his forces have sailed through and his opponent answers in the affirmative if his forces have also been in any of the zones. When this happens the blockader names the days on which he passed through and if they coincide with the enemy's being there his opponent acknowledges and then they quiz regarding grid square, and perhaps time of day. If it transpires that enemy forces were in sight of each other on one day of the preceding week they are laid down on the floor and the battle fought and their subsequent action altered from that previously noted on the map to conform to the results of the battle. The system works well for blockade games, or *guerre de course* campaigns, or for any other map operation in which a greatly superior force is searching for an elusive smaller force, because although

The first raiding cruiser is sighted by Trident's cruiser squadron in zone Top-Middle on Tuesday afternoon in square H3 and is presumably sunk. Unbeknown to the two players the second raiding cruiser crosses the merchantman's wake (square C10) two days too late.

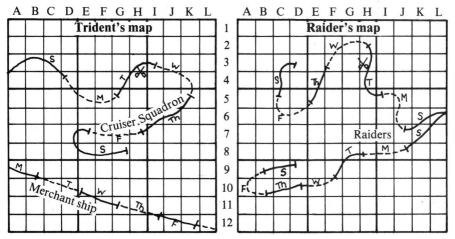

information is often unnecessarily given away by calling out positions it is usually quite vague and it is the blockader or hunter who usually gives himself away, rather than his opponent, who really needs the secrecy.

By far the best method of simulating war at sea with all its mysteries and uncertainties is to use an independent umpire, or more accurately, game controller, with a master map on which he plots the movements of both sides in accordance with instructions given to him by the players. In using this method submarines, ships, disguised raiders, minefields and the like can be employed with all the secrecy essential to their success and, because the umpire can feed the players with only such information or half-information as he thinks they should have, the campaign takes on a realism impossible with the other methods of operation. The system is described in the chapter on the Battle of Texel, where it worked very well. That map campaign only lasted an afternoon and an evening, though, and so avoided the main difficulties of employing an umpire, that of his not being available when the players are raring to fight and vice versa, or of his losing interest just as the campaign is coming to its climax and running away to join a slot-car club instead.

Weather

You can often ignore the complication of weather in wargaming map operations, but in certain types of campaign it can be of crucial importance. When one fleet is blockaded in harbour by a superior force gales which send the gaolers scurrying for shelter present a useful opportunity of escape, whilst the initially successful invasion can turn to disaster if supplies and reinforcements are held up by rough weather or windless calm.

Quite what provision you make for weather changes will depend on the nature of your campaign, but here is a set of rules which you can use, or amend, or chop and change, or ignore as the fancy takes you.

General

  1 One wind direction and one weather condition shall prevail over the whole of the war zone, however large.

  2 Wind direction and weather conditions will be checked for change daily.

Wind

1 Initially the wind direction is chosen as described on page 35. Thereafter a single decimal dice is thrown each day:

1	wind veers 135 degrees
2	wind veers 90 degrees
3, 4	wind veers 45 degrees
5, 6	wind direction unchanged
7, 8	wind backs 45 degrees
9	wind backs 90 degrees
(1) 0	wind backs 135 degrees

2 Initially throw two decimal dice to determine the wind strength.

	Winter	Autumn and Spring	Summer
Calm	0–5 per cent	0–10 per cent	0–20 per cent
Light	6–20 per cent	11–30 per cent	21–50 per cent
Moderate	21–65 per cent	31–75 per cent	51–80 per cent
Fresh	66–90 per cent	76–93 per cent	81–95 per cent

table contd:

	Winter	**Autumn and Spring**	**Summer**
Gale	91–98 per cent	94–98 per cent	96–100 per cent
Storm	99–100 per cent	99–100 per cent	—

Note the percentage score and thereafter adjust the total with a single dice throw each map day:

1	weather worsens	+ 20 per cent
2	weather worsens	+ 15 per cent
3	weather worsens	+ 10 per cent
4	weather worsens	+ 5 per cent
5, 6	weather unchanged	
7	weather improves	–5 per cent
8	weather improves	–10 per cent
9	weather improves	–15 per cent
10	weather improves	–20 per cent

Effect

Ancient period—All galleys at sea in a storm or gale will sink. Damaged galleys may sink in fresh winds—throw one ordinary dice per damaged galley per day and deduct one for every load of water shipped in the battle; 1 or 2 the galley sinks. No rowing in fresh winds.

Napoleonic period—Ships reduced to nil fighting value will automatically sink in a gale, and will sink in a fresh wind if 1 or 2 is thrown on a dice (one dice per ship per map day). Less damaged ships will run the risk of sinking in gales and storms:

$\frac{1}{5}$ damage   1 sinks in a storm

$\frac{2}{5}$ damage   1, 2 sinks in a storm   1 in a gale

$\frac{3}{5}$ damage   1, 2, 3 sinks in a storm   1, 2 in a gale

$\frac{4}{5}$ damage   1, 2, 3, 4 sinks in a storm   1, 2, 3 in a gale

Squadrons and fleets will be dispersed during gales, and storms and will take as long to re-assemble afterwards, as the gale or storm has blown.

If ships are within ten miles of a lee shore in a gale, or 20 miles in a storm, they may be blown on shore and wrecked. Throw one dice per ship per day and 1 or 2 indicates shipwreck. Deduct 2 from the dice for each mast lost, or deduct 1 for each jury mast in use.

Ironclad period—Monitors, and other low freeboard ships must reduce speed to a maximum of two thirds in fresh winds, one third in gales and must heave-to in storms. High freeboard ships will reduce speed to two thirds in a gale and one third in a storm. Any ship which is sailing with one deck sinkage due to battle damage will sink in a storm and a 1, 2, 3 on a dice will sink it in a gale.

Ships must move under steam alone in gales and storms and squadrons will be scattered (see above).

First World War—Destroyers will be reduced to half-speed in gales and must heave-to in storms. All larger ships will reduce speed to three quarters in gales and half-speed in storms, whilst submarines must submerge in storms.

Destroyers suffering 50 per cent damage, cruisers suffering 67 per cent damage and capital ships suffering 75 per cent damage will sink in storms if 1, 2, 3 is thrown on a dice.

Sheltered anchorages

Ships in harbours and sheltered anchorages are protected from the effects of gales and storms, but no repairs to ships afloat, or any coaling may take place whilst strong winds blow.

Repair of battle damage

Using the Dreadnought period as an example, I say that it takes two days' work to repair each points' damage to capital ships and ten days to repair each points' damage on destroyers and cruisers. Any ship suffering less than 50 per cent damage can be repaired in regular anchorages by means of Cofferdams and so on, but more than 50 per cent damage and the ship must be taken to a dockyard with a suitable-sized dry-dock. Gun turrets of capital ships must be repaired in dockyards and take 20 days to repair, although this can be done simultaneously with repairs to hull.

Campaign ideas

The most important feature of any naval wargames' campaign is what we might call the plot. It is a common failing amongst wargamers organising a campaign for the first time to put a lot of work into making splendidly authentic-looking maps and charts, and devising ingeniously complicated rules whilst ignoring the *casus belli* and strategic background to the war.

Our earliest campaigns were of this sort. Invariably two mythical islands lay in an unnamed sea full of compass roses, scales flanked by recumbent mermaids and pictures of galleons and blowing whales. There were no land borders to give rise to dispute, no armies to give material form to their mutual antipathy, and no sea-borne trade to cause rivalry. All we had were two powerful fleets. In such circumstances it is highly unlikely that two real states would have gone to war at all, and even if they had the fleets, having no purpose in life, these would simply have swung at their moorings throughout the phoney hostilities.

Wargames' Admirals and wargames' navies, though, do not bear the weight of great responsibilities on their shoulders and so are more pugnacious than their real-life counterparts can afford to be. Our fleets weighed anchor the moment war was declared, sailed to mid-ocean and fought to the death in one great cataclysmic battle. Wars were thus over in a matter of hours or days, with the victor no more prosperous or secure for winning and the vanquished, except for the loss of his unnecessary navy, no worse-off for losing.

These early campaigns were very unsatisfactory but it was some time before the simple truth dawned on us that, except for one or two prestige fleets in history, expensive things like navies do not simply exist to burden the taxpayer. They exist to protect a nation's vital interests on the seas; trade, colonial links, vulnerable coast lines, etc, and these interests were missing from our campaign.

Nowadays we give less thought to beautifying our maps and more to giving our campaign a realistic commercial or strategic background, with much better results. This does not mean that they are always stimulating, fast-moving affairs—naval warfare invariably involved long periods of tedious waiting and our wargames can occasionally be too realistic in this respect—but usually they provide the players with a sequence of battles and dilemmas which, if not totally like war, are sufficiently exciting for the stay-at-home warrior.

The various roles of wartime navies which should be built into a campaign, either singly or in combination, can be summarised as follows:

1   Protection of friendly sea-borne commerce.
2   Interruption of enemy trade.
3   Landing armies of invasion and keeping them supplied.
4   Protecting the coastline from enemy attack.
5   Maintaining communications with overseas colonies.

To this list could be added such activities as the suppression of piracy, or policing the world with gun boats, but fascinating though these small wars are to read about I have rarely succeeded in creating a worthwhile campaign out of them.

You can set the scene of your campaign by simply pulling them out of your imagination but I often find that the most enjoyable ones have been based, however loosely, on some historical operation. The easiest way is to take the historic situation and reduce it in scale whilst retaining the essential features. The Spanish Armada, for example, might be fought with two fleets of about a dozen ships each, but correct in relative strength and in proportion of ship types. The Spaniards would have the same objectives of rendezvousing with Palma's army and then invading England, and the English would have their same objectives of stopping them in the short period of time for which their stores could be expected to last (and with a very inadequate land army to fall back on).

We do not always have model ships of the correct period to refight an operation which interests us, but it is usually possible to transfer the essential features of a campaign to another historical period. The background to the Battle of Tsu-Shima, for instance, is quite similar to the Armada, but using steam battleships.

The most fascinating campaigns to refight are those might-have-beens which were planned or proposed but never put into execution. The First World War presents a useful number of these unrealised operations and we have fought some excellent campaigns trying to capture Heligoland or Borkum as per Jackson, or link up with the Russians in the Baltic as per Lord Fisher. All these campaigns eventually resulted in costly failures, which was a pleasant vindication of those doubters at the Admiralty who we had previously written off as shortsighted because they had blocked the projects as impractical.

Here are a few of the campaigns which we have fought with success over the past few years; they may give you some ideas for your own map operations.

The Siege of Hagage
(loosely based on the Siege of Carthage)

The besieged city, with its harbour, was made of card and balsa and permanently set up on the wargames table manned with 15 mm model soldiers. The city at A was garrisoned by about 100 Greeks and besieged by about 250 Romans using mangonels, siege towers, etc. A fairly complex system of supply and consumption was employed by which besiegers and besieged ate up supply points and fell in fighting ability and activity as shortages caused them to go hungry. The Romans were supplied from their base city B, one day's journey away, and the Greek garrison was supplied from their base at C, two and a half days away. Each side had a fleet of ten war galleys and ten sailing

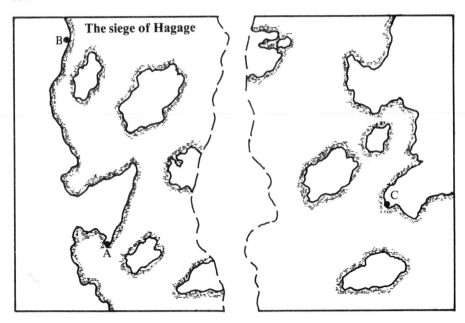

merchant ships with which to keep their land forces supplied, prevent supplies reaching the enemy and mount raids. If the city held out for six weeks it was assumed that a relieving army would arrive and raise the siege, but as it happened, during the fourth week the Roman transport fleet, sailing with only a two-galley escort, was annihilated by the Greek fleet and eventually hunger caused them to abandon the siege.

The War in Ireland, 1702

This naval campaign was fought in conjunction with two other wargamers who were fighting a land campaign in Ireland. The land and the sea campaigns were played more or less independently, except that the times and dates were synchronised and the supplies and reinforcements which the Generals received depended upon the success of their naval colleagues. Supplies had to be landed at the ports selected by the Generals as their bases; distribution to the field armies by wagon train and barge was a matter for the land wargamers. If the soldiers were inadequately supplied their fighting and marching abilities diminished until they eventually starved. Initially an English Army of about 150 men was in Ireland. About May 1 (exact date pulled out of a hat) the French Army of invasion of about 300 would sail from Brest, in two waves as there were not enough ships to carry the whole army in one go.

About May 21 an allied Dutch Army was picked up from Amsterdam and taken to Ireland in English ships. About June 14 France's Bavarian allies were assembled at Brest for transportation, and about July 5 Danish and Prussian mercenaries in English pay assembled at Ostend. Each side had 12 men-of-war and 12 merchantmen which as they suffered losses made troop-carrying and supply deliveries more and more difficult to achieve, which resulted in more than one inter-service row. In the event, the French Fleet performed outstandingly but their army was soundly beaten on land and had to be evacuated.

An 18th century trade war

 The cities A–F are neutral commercial ports and the two countries (shaded) are trading nations at war with each other. Each starts the war with ten ships of the line, four frigates and 25 merchant ships. Each port deals in a different commodity (taking the form of counters marked A–F) and a merchant ship may carry one unit (counter) of goods. A full set of six units brought home by the merchant fleet is worth £1,000 revenue to that country and five of the six units is worth £500; less than five of a set is worth nothing. At the end of each month the two players total up the amount of revenue earned and then pay out the wage bill of their warships at the rate of £20 per gun. Any ship which cannot be paid off must be brought home and laid up until the next month end when, if funds are sufficient, it may be re-commissioned. New warships may be purchased at a cost of £50 per gun for delivery two months after the sum has been paid and merchant ships may be purchased for £800 for immediate delivery. The winner is the player who still has warships in commission when the enemy's are all sunk or laid up.

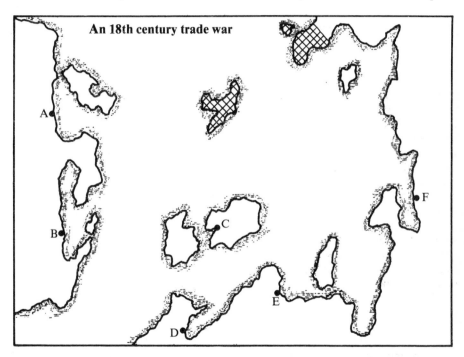

An 18th century trade war

War in the Far East, 1880
(loosely based on the Russo-Japanese War)

 This campaign was fought as the supply operation following the invasion of a peninsula, but no model soldiers were used. The French had a force of ten Ironclads operating from any of three fortified harbours on the Pong-Ho peninsula. The British had a force of roughly double strength, plus supply ships with which they had to bring war stores from their base 500 miles away to a fourth port, on the tip of the peninsula, which they had previously captured. If the British brought 200 supply points into the port in one week they would advance one zone, or two zones if they received 300. If supplies delivered were

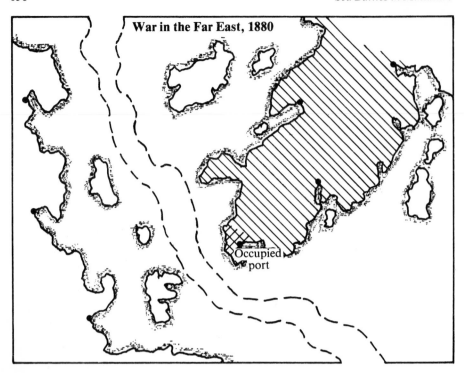

less than 100 they retreated a zone and if less than 50 they retreated two zones. As they advanced zone by zone up the peninsula, they would capture French ports, with a consequent reduction in enemy coaling and repair facilities and an increase in the number of docks available for unloading supplies. If the French pushed the British off the tip of the peninsula, of if the British captured the last port town, that side would be deemed to have won. The French were able to limit supplies to the British Army by shelling their harbours at bases where supplies were accumulated for loading (one point per two small or one large shell hit), by sinking merchant ships loaded or unloaded, or by forcing the merchant ships to sail in large armed convoys, thus causing long delays as the harbours could only load/unload two ships at a time.

The Baltic Project—First World War

A more flexible system of land strategy than that described above was used in this campaign. Cardboard counters, representing divisions, were moved about the land map at a maximum speed of eight miles per day. Each division commenced with a fighting value of six points which was added to the score of a dice throw whenever forces met in combat. The force with the superior total compelled the enemy to retire, with the loss of one, two or three points depending upon the degree of superiority. Supplies were delivered by sea to ports (captured by the armies) and distributed by railway. A division within ten miles of a railway line connected to a supply port was supplied, providing that the port had an adequate quantity for all the divisions drawing from it and providing that the rail link was not cut by an enemy force, otherwise its fighting value was reduced.

The campaign was fought by three players, British, Russian and German, over two maps, one showing the sea, and a larger-scale map showing North Germany in detail. Relative strengths of forces were: Britain, navy 4, army 1; Russia, navy 1, army 4; Germany, navy $2\frac{1}{2}$, army 3. The Russians had inadequate transport to ship their army to the North German coast and to supply it there. The objective of the campaign was for the Russians to carry out such an invasion with assistance from the British Army and Navy which had to break through the Skagerrak and Kattegat, and still maintain contact with home to be effective. Capture of six nominated towns, reckoned to be centres of industrial production, would signify the end of the war. In the event mines and submarines in the narrows reduced allied naval superiority to such an extent that the armies could not be adequately supplied, and so were defeated.

The First World War in the North Sea

It was an early ambition of ours completely to re-create the war in the North Sea between 1914 and 1918, but the problems and complexities of running a map operation lasting four years, even though only in game time, defeated us. As a compromise we devised a series of five one-off map-based wargames, controlled by an umpire, such as that described in Chapter 11. About a dozen objectives were written down and pulled out of a hat at the beginning of each game. The games were independent of each other, except that they were consecutively set in 1914, 1915, 1916, 1917 and 1918, with the two fleets at the strengths which they had in those years. Newly-completed ships would join the fleet prior to the appropriate year's game but any ships sunk would not re-appear and ships suffering 50 per cent damage would miss the next battle but appear in the one after that. The advantage of waging the war in this way was that it gave the Germans the opportunity to hammer the Grand Fleet early in the war when the British capital ship superiority was slight, without becoming burdened with a too-lengthy campaign.

Appendices

1 Some sources of supply of model ships

This list is by no means exhaustive but it will give some idea of who the principal suppliers are and what they sell. The advertisements in the magazine *Military Modelling* are well worth scanning for news of new products.

Airfix Ltd (generally available): A small range of 1:1,200-scale plastic kits of Second World War ships. A small range of small-scale period ships suitable for conversion.

Fleetline Model Co, 2 Lees Lane, Gosport, Hants: 1:1,200-scale Second World War and modern ships.

Heroics and Ros Figures, 36 Kennington Road, London SE1: 1:1,200-scale galleys.

Leicester Micro Models Ltd (see Trafalgar Models).

Navwar, 48 East View, Barnet, Herts, EN5 5TN: Enormous range of 1:3,000-scale metal models 1890s to end of the Second World War. Ensign 1:1,200-scale metal Second World War ships. 1:1,200-scale resin and metal Napoleonic warships. Resin and metal 1:1,200-scale American Civil War Ironclads. 1:2,400-scale Second World War ships. Rules, dice, etc.

Skytrex Ltd, 39 Ashby Road, Loughborough, Leicestershire: Enormous range of 1:3,000-scale metal pre-Dreadnought, First and Second World War models. Also rules, dice, etc.

Starcast Miniatures, 140a Morley Road, London: Starcast 1:2,400-scale metal First and Second World War warships. Micro-Matrix 1:1,000-scale Napoleonic warships. 1:450-scale ancient galleys.

Trafalgar Models, 122 Lazy Hill Road, Aldridge, Walsall, West Midlands: now suppliers for Leicester Micro Models in addition to own products—1:4,800- and 1:3,000-scale metal First and Second World War warships; 1:1,200-scale smaller vessels of the Second World War.

2 Some useful further reading

On Ships

Battleships and Battle Cruisers 1905–1970, by Siegfried Breyer; Macdonald.
Battleships of World War 1, by Antony Preston; Arms & Armour Press.
Battleships 1856–1977, by Antony Preston & John Batchelor; Phoebus.

British Battleships 1860–1950, by Dr O. Parkes; Seeley, Service & Co.
British Destroyer 1892–1953, by E.J. March; Seeley, Service & Co.
French Warships of World War 1, by Jean Labayle Couhat; Ian Allan.
Jane's Fighting Ships, Reprints for the years 1898, 1905, 1914, 1919 and 1939; David & Charles.
Man-of-War, by D. MacIntyre and B.W. Bathe; Jupiter Books Ltd.
Metal Fighting Ships of the Royal Navy, by E.H.H. Archibald; Blandford.
Modern History of Warships, by Wm. Hovgaard; Conway Maritime Press.
Oared Fighting Ships, by R.C. Anderson; Argus.
Sailing Ships and Sailing Craft, by George Goldsmith-Carter; Hamlyn.
The Ship, by Bjørn Landstrøm; Allen & Unwin.
Submarines of World War 2, by Erminio Bagnasio; Arms & Armour Press.
Warships from 1860 to the present day, by H.T. Lenton; Hamlyn.
Warships of World War 1, by H.M. Le Fleming; Ian Allan [British & German only].
Warships of World War 2, by H.T. Lenton and J.J. Colledge; Ian Allan [all powers].
Wooden Fighting Ships of the Royal Navy, by E.H.H. Archibald; Blandford.
World Warships in Review 1860–1906, by John Leather; Macdonald and Jane.

On naval warfare

The Ancient Mariners, by Lionel Casson; Victor Gollancz.
The Battleship Era, by Peter Padfield; Pan Books.
The Floating Bulwark, by Douglas G. Browne; Cassel.
From Dreadnought to Scapa Flow, by A.J. Marder; Oxford University Press.
A Guide to Naval Strategy, by B. Brodie; Princeton University Press.
Gunpowder and Galleys, by I.F. Guilmartin.
A History of Russian and Soviet Sea Power, by D.W. Mitchell; Andre Deutsch.
The Influence of Sea Power upon History, by Alfred Thayer Mahan; Methuen & Co.
The Roman Art of War under the Republic, by F.A. Adcock; W. Heffer & Sons.
Sea Power a Naval History, by E.B. Potter & C.W. Nimitz; Prentice Hall International Inc.
The Strategy of Sea Power, by S.W. Roskill; Collins.
25 Centuries of Sea Warfare, by Jacques Mordal; Souvenir Press.

On naval wargames

Naval Wargames, by Donald F. Featherstone; Stanley & Paul.
Naval Wargames—World Wars 1 and 2, by Barry J. Carter; David & Charles.
Sea Battle Games, by P. Dunn; Model and Allied Publications.

Naval wargames rules

American Civil War Ironclads, Navwar.
Coastal Warfare World War 2, Navwar.
Greek Naval, Wargames Research Group.

Greek Naval Warfare, London Wargame Section.
Napoleonic Naval Warfare, Navwar.
World War 1 Naval, Skytrex.
World War 2 Naval, Leicester.